Nobel Genius

# NOBEL GENIUS

## Prizes, Prestige and Scientific Practice

Edited by

Nils Hansson & Ad Maas

Leiden University Press

Cover design: Andre Klijsen
Cover illustration: Engraving in memorial of Alfred Nobel. Located in the City Hall of Stockholm.
© Kristoffer Pettersson.
Lay-out: Crius Group



ISBN 9789087284138
e-ISBN 9789400604582 (e-PDF)
https://doi.org//10.24415/9789087284138
NUR 740



Printed and bound by CPI Group (UK) Ltd, Croydon, CR0 4YY

# Table of Contents

# Prizes? What prizes?

*Preface by Klaas Landsman*

Until recently, my view of scientific prizes (including memberships of Royal Societies and Academies etc.) was that really great scientists like Einstein do not need them since they have "made it" anyway, that really bad scientists should not get them either for obvious reasons, and that in the remaining grey zone prizes do more harm than good, since for every scientist who wins a prize there are dozens who do not, although they would be just as deserving of them. Still in the grey zone, even if there actually are tiny differences between winners and non-winners, these tend to be amplified by the Matthew effect (i.e., those who win one prize have a much better chance of winning the next even without any further achievements), which amplifies an already slightly unfair situation to sometimes colossal proportions. Furthermore, in the spirit of Michel Foucault, I felt that awards often reflect power structures and nepotism in that winners or jury members who choose them may not be the best scientists but are those that rule science: this applies from best poster awards for youngsters to Nobel Prizes – just think of the many women who should have won the latter but were sadly passed over in favour of men – and it even applies more abstractly to dominant areas of research. Finally, as in (top) sports, prizes increase the danger that doing (top) science becomes a race whose goal is winning, as opposed to lofty ideals such as beauty and truth, or simply self-realization through excellence as such. In other words, prizes promote competition rather than collaboration.

My negative attitude changed considerably after I won a prestigious prize myself (and a few years earlier had become a member of the Royal Netherlands Academy of Arts and Science). Three recent Nobel Prizes also played a role in doing my about-face: to Roger Penrose in 2020 for Physics (robustness of black hole formation), to Guido Imbens in 2021 for Economic Sciences (analysis of causal relationships), and to Katalin Karikó in 2023 for Physiology or Medicine (mRNA vaccines against COVID-19). Each story is one of passionate truth-finding from an outsider point of view, without any obvious lobby or power structure behind the award, which recognized work that genuinely benefits mankind – including Penrose's, since besides a fairer and healthier world, understanding the Cosmos is also one of our goals. In a rather more infinitesimal realm of importance, I was very pleased to learn that other people even knew my work, in addition to appreciating it. And surely many of those who are genuinely surprised by some prize feel likewise.

In conclusion, I still believe that all my earlier arguments are correct and would not shed a tear if all scientific prizes were abolished (having cashed mine in already); but I am also beginning to see a glimpse of light. It is popular these days to emphasize the value of team science, but in many cases public (and monetary) recognition of someone's individual struggle and ensuing insights and achievements are at least as important. Well-deserved prizes may serve this goal at any career stage.

*Klaas Landsman is Professor of Mathematical Physics at the Radboud University Nijmegen. In 2022 he received the Spinoza Prize, the highest award in Dutch science.*

CHAPTER ONE

# Introducing Prize Studies: Perspectives on Reward Mechanisms in Science

*Nils Hansson & Ad Maas*

Almost every researcher's curriculum vitae follows the same general template: lists of scientific publications are followed by grants and memberships in national or international societies. In addition, scholars may finally add prizes. Unlike the impact of publications and citation patterns, and grants, which can be quantified precisely, prizes seem to elude clear classification in scientific evaluation systems. In other words: do prizes matter? Some view them merely as the cherry on the cake, but not crucial for reverberation or career advancement.[1] Others see them as atavisms: Why praise individual scholars when we know that research is often pursued in large multidisciplinary teams, sometimes in collaboration with partners from outside academia?[2]

With this book, we wish to outline a view on the role of prizes in science from a historic perspective and offer conclusions that will be relevant for universities, research centers and academic promotion committees in evaluating academic reputation. We see prizes in academia as a barometer of how we look at scientific excellence and how we value scientific practices and results, both within and outside the scientific community. Indeed, in contrast to lists of publications and grants, prizes also appeal to broader audiences – most of all, of course, Nobel Prizes. Following the early October announcements, international media outlets and local newspapers all over the world cover the laureates. Studying prize cultures therefore can help to unravel social aspects of both the scientific practice and how science is regarded in society. In a nutshell, the book aims at establishing **prize studies** as a field of research in its own right.

Science prizes shape careers, make research visible for the public, and create role models in academia and beyond. They provide evidence of scientific recognition, prestige and credit, and play a key role in the evaluation of individual scientists, for example in academic promotion committees. Nevertheless, whereas the alleged impact of citations and external funding has been a growing field of international research for decades in science studies and beyond, the understanding of prize cultures and their dynamics, in different countries or in specific fields, has remained surprisingly superficial. According to Pierre Bourdieu (1930-2002),[3] literary prizes contribute to the conversion of symbolic capital into economic capital,[4] but it is questionable if this also is the case in the natural sciences and medicine. Prize studies then should explore the conditions, processes, functions and consequences of scientific awards. They could contribute to modern approaches in history and sociology of science that focus in particular on the social context of scientific practices and give new insights into scientific reward mechanisms and the role of credit, status and impact in academia.

We seek to show in this volume that the study of scientific prizes as a *pars pro toto* can provide important perspectives on scientific practices and cultures, both within the scientific community and in broader society. Prizes are about status, and status is a sensitive indicator of social relations. Within the scientific community numerous social mechanisms are involved in prizes. What does winning important prizes mean for a scientist's career? How do prizes privilege certain individuals, nationalities, fields and research interests, and marginalize others? What social mechanisms are involved in the selection of candidates and winners? And what does this all say about the scientific community? Which are the effects of prizes for individual scientists, institutions, scientific communities? To what extent do these effects depend on individual dispositions like gender, origin, age, and so forth?

Highly reputed prizes are used to boost scientific work, with prestigious awards being handed out to academics who have conducted and published research judged as important, outstanding, or to use the words of Alfred Nobel, that has "conferred the greatest benefit to humankind" (Nobel's will of 1895).

Several chapters in this volume reveal the pervasiveness of the narrative of the "genius," or "hero" of science in the history of the Nobel Prizes. This persona has been inherently connected to the very idea behind the Nobel Prize, which after all singles out individuals and awards personal achievements, and it has also fashioned the public image of Nobel laureates. As Källstrand succinctly writes in his chapter: "The Nobel prize has become an institution with the power to routinely create scientific heroes." This narrative therefore is a leitmotiv in this volume.[5]

The focus on brilliant individuals, in the meanwhile, has not remained unchallenged. It has been questioned, for instance, by historians and sociologists of science who consider scientists and their insights (however brilliant) rather as a product of their times, or emphasize the collaborative nature of science. Especially in recent times of big science, the relation between achievement and the individual has been increasingly questioned.

We wish to analyze the impact of prizes from three perspectives. First, from the perspective of an institution that awards the most prestigious prize in the sciences: the Nobel Foundation, as well as from Nobel laureates. Second, from the view of the scientific community, and third, from the perspective of the public understanding of science. The chosen foci of investigation are suitable for pursuing the research questions in a methodologically exemplary manner and to prepare the ground for further research.

### State of research on prizes

Following the lead of literature on scientific merit, we do not view excellence as a self-evident category. There is no way to measure scientific brilliance objectively. For example, with regards to prizes, it is not possible to compare medals in science with medals in sports. The results of athletes (who runs the fastest, who jumps the highest) can be determined objectively, but this is not the case in academia. In addition, participants in the Olympic Games compete against others in the same niche (hurdles, long jump, etc.), while the performances of scientists are hardly comparable. This makes scientific competition dependent on other factors, not least marketing strategies and personal networks. In other words, the entire prize decision-making process is an example of the staging of excellence. Thus, we consider prizes not only as the result of a strong performance, but also as a result of a construction or an attribution process.

The ongoing debate on the attribution of scientific excellence has so far centered on the distribution of research funding.[6] In his book *Die akademische Elite*,[7] sociologist Richard Münch goes into more detail about the reputation of researchers in the review process of externally funded projects. Such accounts about the task to single out the most promising scientists and projects have sparked current debates on how to find new, unorthodox ways of selecting the most important projects. For example, the Health Research Council of New Zealand launched a new type of "lottery grant" to fund research. Along those lines, Swiss scientists suggested to toss a coin to decide which candidates to appoint to academic positions. The idea is that incorporating randomness reduces bias and promotes diversity among the applicants. To our knowledge, this strategy has not yet been used in prize competitions. Prizes still benefit from an aura of objectivity, which needs to be understood more in detail.

Reputation and recognition in their turn rely heavily on visibility of scientists. Prizes bring exactly that: they give scientists status by making them visible, and their visibility gives status to prizes. To have these effects, prizes need to be staged, performed, and celebrated.[8] In the context of this book we understand attributing excellence to someone as the performative act of ascribing prize-worthiness to a person based on the perceived quality of his or her scientific work.[9] Staging is again understood as the method and process of presenting, framing, and contextualizing institutions, people, and facts.[10]

Earlier work has reflected on the impact of prizes in different fields and what the "prize population" has looked like.[11] Ma and Uzzi listed 3,000 scientific prizes in different disciplines and tried to reconstruct the careers of more than 10,000 prizewinners.[12] Among the winners since the turn of the millennium, female scientists are significantly underrepresented.[13] Scholars have also raised the issue of how prizes have been related to academic productivity,[14] stating that winners' productivity declines after the award year of big prizes.

No award has inspired as much research as the Nobel Prize, focusing mostly on the laureates and their work. Harriet Zuckerman's book *Scientific Elite: Nobel Laureates in the United States*,[15] based on interviews with U.S. laureates, is an intriguing analysis of Nobel laureates to date. What the previous research shows, we conclude, is the need for a comprehensive approach, looking at prizes as a field of study. The research field needs to be developed with studies on the significance of prizes, which understand the different perspectives of prize-awarding institutions, prize-winners, the scientific community and the public as a coherent research complex.

The Nobel Prize, in addition to serving as the most influential certification of extraordinary achievement, constitutes a lens through which we can examine some aspects of the scientific enterprise – from the broader role of credit and recognition in academia to the role of science and scientific leaders in contemporary society, or as Jeffrey Flier, former dean of Harvard Medical School, put it: "Though Nobel Prizes inhabit the extreme end of the prize spectrum, and therefore have many unique attributes, Nobel effects ripple like a force field through the broader ecosystem of science, illuminating general principles of scientific credit and warning of potential future problems in this area."[16]

This volume focuses on the most famous (if not notorious) of all prizes in the sciences: the Nobel Prize. These prizes are the ideal example for investigating the mechanisms of scientific recognition (university ranking companies also use them as a parameter). Like no other prize, studying the history of Nobel Prizes reveals the impact prizes can have on the scientific practice and can furthermore help to unravel insights in the public understanding of science. How do Nobel Prize laureates use their position both within the scientific community and as public figures to fashion a certain image of themselves?[17] And what does that say about the public understanding and appreciation of science? The discussions in this volume on the narrative on the geniuses or heroes of science, which has largely underpinned the public perception of Nobel Prize laureates, provide a contribution to the historiography on scholarly personae that has been developed in recent years.[18] Our exploration is based on several examples from Europe in general and from the Netherlands in particular, a country which received relatively many Nobel Prizes during the first decades of their existence and therefore forms an interesting case for exploring social aspects connected to it.

## Introducing the chapters

The contributions in this volume focus on the role of Nobel Prizes in Dutch and western history of medicine and science. With this varied collection we aim to explore and delineate the field of "prize studies" in history of science. So far, our understanding of how Nobel Prizes and other major awards in the sciences have been used as a symbol for excellence has remained poor. Whereas historians dedicated themselves increasingly to social aspects of science and marginalized groups, historical interest in Nobel Prizes, together with the demise of "old-fashioned" great-man historiography, apparently have been largely disappearing.

The papers form a rich diversity of contributions that all address one or more of the topics mentioned above, and further our understanding of social aspects of science. They critically discuss ideas of authority, impact and scientific heroism in general, and relate to the social organization of the sciences, to originality, feelings, and creativity in medicine.

Källstrand traces the roots of the hero-of-science-narrative in his chapter and offers a discussion on the relevance of the Nobel Prize based on the effect that controversial choices and controversial Nobel laureates have had on the prize status. Throughout the twentieth century, Källstrand notes, the high status of the Nobel Prize has prevented controversies doing harm to the prize. He argues, however, that this is changing. The status depended to a high degree on the cultural status of celebrating "great men" for their discoveries – and this status has come increasingly under scrutiny over the last few decades. Thus, rather than individual decisions or laureates being seen as inappropriate, it is the underrepresentation of women that can undermine the prize's status.

In their chapter, Ad Maas and Louise Lagarde regard the public perception of Nobel Prize laureates through the lens of artefacts connected to Nobel Prize laureates in a museum collection, that is, that of Rijksmuseum Boerhaave, the Dutch National Museum of the History of Science and Medicine in the Netherlands, which opened its doors in 1931. For a long time, as Maas and Lagarde detail, the way these objects were presented was in line with a somewhat romantic heroes-of-science-image that was common in academic circles, where Nobel Prize laureates were regarded with reverence. Keeping and displaying their artefacts as "relics" – a word the first museum-director used – was part of this veneration. From the 1970s onwards this view was replaced by a more detached and intellectualistic approach in which the focus came on scientific instruments.

The "heroes" of the Leiden Museum had always been regarded as brilliant scientists who were *also* Nobel Prize laureates. Only after the turn of the century, did the Nobel Prize come to the fore, and were the scientists presented as such, Nobel laureates, in the first place. In this era of mass-communication and marketing,

"Nobel" proved to be a perfect "brand" to present the scientists to broader audiences. In recent years, Maas and Lagarde conclude, the call for more "diversity" and "inclusivity" in museums has posed a new challenge to presenting the Dutch Nobel Prize laureates, all white males, in a suitable manner.

The way Nobel Prize laureates have been presented in literature is strikingly different from that in museums, Daniela Link's chapter reveals. Link engages with the representation of Nobel Prizes and Nobel Prize laureates as motifs in literature. Also in the novels she addresses, the genius-narrative plays a significant role, albeit often in a perverted way: rather than as altruistic benefactors of mankind the protagonists of the books in question appear as egocentric and calculating (if not cynical) personas full of personal shortcomings: "The Nobel-biotope in (especially recent) literature," Link writes, "is inhabited with neurodivergent figures, the solitary and socially deviant geniuses." In literature, the lofty Nobel Prize then often acts as a foil to express the human deficiencies of the main characters.

"The literary genius of science," Link explains, "draws on the archetypal hero, who behaves as symbol of morally good action and savior in distress," with a willingness to self-sacrifice. Characteristically, this literary hero is inclined to follow his own path to achieve his goal: "Non-conformism and dissociation can thus be identified as signs of heroism." This specific trait of the hero as outsider got a distinct manifestation in the persona of the genius around 1900, when the burgeoning mass-society informed a longing for individuals who elevated themselves figuratively above the masses. But the genius as eccentric loner also gave rise to the persona of the power-hungry, mad genius, and it is this guise that has particulary resonated in literature.

Midway through the twentieth century, Annelie Drakman shows in her chapter, a variant of the genius-image emerged that emphasized playfulness and fun as scientific virtues. Up to this point, why one pursued science had primarily been motivated either by its practical usefulness or by the engendering of awe and wonder. But, after World War II, scientists began speaking publicly and emphatically of their own enjoyment of science. Why did it become acceptable to say that one did science for the fun of it? To answer this question, self-depictions by two of the most influential public intellectuals of the time are investigated: Richard Feynman (Nobel laureate in Physics, 1965) and James Watson (Nobel laureate in Physiology or Medicine, 1962). Drakman argues that depicting oneself as having fun was a way of displaying vitality, by means of showing off characteristics and traits such as impulsivity, charm, creativity.

Whereas the image of the hero-of-science prevailed in the public perception almost unabatedly throughout the twentieth century, it was dismantled in professional circles of philosophers, sociologists and historians of science. Jelmer Heeren in his chapter details this demise by using the work of the prominent

Dutch historian of science Reijer Hooykaas as a case study. Heeren sets out that, when Hooykaas started his career in the 1920s, the hero-of-science-narrative was underpinned by the then dominant, positivistic "Whiggish" view on the past. This approach divided the history of science in scientifically open-minded observers, the "heroes" who contributed knowledge we now know to be true, on the one hand, and superstitious and dogmatic speculators who had it wrong, on the other.

Hooykaas belonged to the first generation of historians of science to dismiss this positivistic narrative. Already in the 1930s, he downplayed the alleged revolutionary nature of the work of Antoine Lavoisier, the towering figure in the history of chemistry. It was, according to Hooykaas, the "humble" task of the historian to regard historical actors in the intellectual context of their time rather than singling them out as isolated geniuses. Hooykaas predated the work of post-positivists like Thomas Kuhn, but disagreed with historians of science who dismissed the singular role of the individual scientist altogether.

All the papers in this volume so far bear witness to the huge public appeal of the Nobel Prizes. In his chapter on Emil Behring, the first Nobel Prize laureate in Physiology or Medicine, however, Christiaan Engberts addresses a case of a Nobel Prize that instead scarcely evoked any response. His biographers at best mentioned it as a minor footnote in his career. Engberts dismisses the argument that, in the early days, when Behring got his prize, the Nobel Prize's standing had yet to be established. He instead points to circumstances both in the professional sphere and in the public perception of Behring's work that contributed to the silence surrounding his Nobel Prize.

As to the first, Engberts sets out that maintaining cordial relations with his peers in Germany, did not belong to Behring's strongest qualities, to say the least. After alienating himself from his colleagues with his bellicose nature, they were not really inclined to put him on a pedestal at the occasion of his Nobel Prize. In striking contrast to his reputation among his colleagues was his enormous status in the public domain, where the inventor of the life-saving serum against diphtheria was hailed as "the benefactor of humanity" and "the savior of children." Engberts claims that the Nobel Prize simply had little to add to Behring's "already stellar reputation" as a civic hero. Behring's reputation, in sum, was too low within, and too high outside academic circles. "Apparently," Engberts concludes, "public and academic ideals of good scholarship did not overlap entirely."

Such academic ideals of good scholarship are further elaborated in the subsequent chapters, in which Nobel Prizes serve particularly to expose social dynamics within the scientific community. Thus, Rob van den Berg expands on the seemingly remarkable decision of the Nobel Committee to grant the Nobel Prize for Physiology or Medicine in 1929 to Christiaan Eijkman, for his contribution to the discovery of what we now call Vitamin B1. After all, Eijkman himself had opposed the correct interpretation for his own experimental results for more than three decades. His former assistant Gerrit Grijns, by contrast, was the very first scientist to point to the "partial hunger" resulting from vitamin deficiency, but his work was largely ignored, both by the Nobel Committee and by Eijkman. Van den Berg's case clearly shows the downside to picking out individuals to be rewarded for the scientific achievement in which many researchers have in fact been involved. Having a status within the scientific community and having access to the right communication-channels then become pivotal to getting the credits one does or does not deserve.

The article by Daniela Angetter deals with the knowledge transfer between the Nobel laureates Konrad Lorenz and Nikolaas Tinbergen. Both have had a significant impact on studies of instinctive behavior in very different ways. However, their different ways of working could be perfectly combined. The article also sheds new light on the general differences and similarities between the researchers, especially regarding their social, scientific, and ethnic backgrounds, for example, in relation to the Second World War and the Nazi regime.

Drawing on nominations gathered in Nobel Prize archives for medicine and for peace, the two final contributions argue that Nobel Prize nominations are unique sources to reconstruct how excellence was defined for individual scholars during the first half of the twentieth century. The first by Nils Hansson, Giacomo Padrini, Andreas Winkelmann and Mathias Schütz pinpoints German anatomists as Nobel Prize candidates. Focusing on Albert von Kölliker, Hans Spemann and Wolfgang Bargmann as examples, they shed light on the reasons for how the trio was portrayed by their nominators, why Kölliker and Bargmann were turned down by the Nobel Committee in Stockholm, and why Spemann's nominators succeeded in convincing the jury in 1935. A review of the Nobel Committee evaluations shows that Kölliker's work was deemed too old, and, in Bargmann's case, other nominated candidates were viewed as more important. The examples of Kölliker, Bargmann, and Spemann shed light on the development of anatomy as a scientific field and its contribution to medical knowledge and practice, as considered by the Nobel Committee. In the final piece, Leander Scheel and Nils Hansson show that physicians were not only nominated for the "Physiology or Medicine" Nobel Prize category. They focus on doctors who were nominated for the peace prize and reconstruct the nomination careers of Louis Lazare Zamenhof, Josué Apolônio de Castro, and Albert Schweitzer.

**Acknowledgement**

We thank all participants at the symposium "Does Science need Heroes?" held at Rijksmuseum Boerhaave in Leiden September 29-30, 2023, for fruitful discussions.

## Notes

1 Katarina Nordqvist and Pauline Mattsson, "Nobel Prize Awarded Discoveries and Commercialization: The Role of the Laureates," in *Attributing Excellence in Medicine: The History of the Nobel Prize*, edited by Nils Hansson *et al.* (Leiden: Brill, 2019), 188–206.

2 Arturo Casadevall and Ferric C. Fang, "Is the Nobel Prize Good for Science?" *FASEB J.* 12 (2013), 4682–4690.

3 Pierre Bourdieu, *Les règles de l'art. Genèse et structure du champ littéraire* (Paris: Seuil, 1992).

4 Gisele Sapiro, "The Symbolic Economy of the Nobel Prize in Literature: How it Counters or Reproduces Modes of Domination," *Poetics* 101 (2023), 101823.

5 Sometimes the "genius" or "hero" of science are used as synonyms, but some of the papers treat them as separate entities and investigate their mutual relationship. See in particular the chapter by Link.

6 Olof Hallonsten, "Introduction to Special Section: Causes and Consequences of the Current Evaluation Regime in (Academic) Science," *Social Science Information* 4(61), (2023) online first, https://doi.org/10.1177/05390184231151610.

7 Richard Münch, *Die akademische Elite* (Frankfurt am Main: Suhrkamp, 2007).

8 Nils Hansson, *Wie man keinen Nobelpreis gewinnt – Die verkannten Genies der Medizingeschichte* (München: Gräfe & Unzer, 2023).

9 Ulrich Bröckling, *Postheroische Helden. Ein Zeitbild* (Berlin: Suhrkamp, 2020).

10 "Beyond the Nobel Prize: Scientific Recognition and Awards in North America since 1900," special issue edited by Nils Hansson and Thomas Schlich, *Notes & Records: The Royal Society Journal of the History of Science* (2024), online first, https://doi.org/10.1098/rsnr.2022.0015.

11 *Historical Studies in the Nobel Archives: The Prizes in Science and Medicine*, edited by Elisabeth Crawford (Tokyo: Universal Academy Press, 2002).

12 Yifang Ma and Brian Uzzi, "Scientific Prize Network Predicts who Pushes the Boundaries of Science," *Proceedings of the National Academy of Sciences USA* 115(50) (2018), 12608-12615.

13 Lokman I. Meho, "The Gender Gap in Highly Prestigious International Research Awards, 2001–2020," *Quantitative Science Studies* 2(2021), 976–989.

14 George J. Borjas and Kirk B. Doran, "Prizes and Productivity: How Winning the Fields Medal Affects Scientific Output," *Journal of Human Resources* 50 (2015), 728–758.

15 Harriet Zuckerman, *Scientific Elite: Nobel Laureates in the United States* (New Brunswick: Transaction Publishers, 1995).

16 Jeffrey S. Flier, "Foreword," in *Attributing Excellence in Medicine: The History of the Nobel Prize*, edited by Nils Hansson, Thorsten Halling and Heiner Fangerau, IX–XI (Leiden: Brill, 2019).

17 "The Consecrating Power of the Nobel Prize in the Global Literary Field," edited by Jørgen Sneis and Carlos Spoerhase, *Poetics* 101 (2023), 101821.

18 See, for instance: *Epistemic Virtues in the Humanities and the Sciences*, edited by Herman Paul and Jeroen van Dongen (Boston: Springer, 2017); Herman Paul, "What is a Scholarly Persona? Ten Theses on Virtues, Skills, and Desires," *History and Theory* 53 (2014), 348–371; Lorraine Daston and H. Otto Sibum, "Introduction: Scientific Personae and Their Histories," *Science in Context* 16 (2003), 1–8.

## Bibliography

Borjas, George J. and Kirk B. Doran. “Prizes and Productivity: How Winning the Fields Medal Affects Scientific Output,” *Journal of Human Resources* 50 (2015): 728–758.

Bourdieu, Pierre. *Les Règles de l’art. Genèse et structure du champ littéraire*. Paris: Seuil, 1992.

Bröckling, Ulrich. *Postheroische Helden. Ein Zeitbild*. Berlin: Suhrkamp, 2020.

Casadevall, Arthuro and Ferric C. Fang. “Is the Nobel Prize good for science?” *FASEB Journal* 12 (2013): 4682–4690.

Crawford, Elisabeth, ed. *Historical Studies in the Nobel Archives: The Prizes in Science and Medicine*. Tokyo: Universal Academy Press, 2002.

Daston, Lorraine and H. Otto Sibum. “Introduction: Scientific Personae and their Histories,” *Science in Context* 16 (2003), 1–8.

Flier Jeffrey S. “Foreword.” In *Attributing Excellence in Medicine: The History of the Nobel Prize,* edited by Nils Hansson, Thorsten Halling and Heiner Fangerau, IX–XI. Leiden: Brill, 2019.

Hallonsten, Olof. “Introduction to Special Section: Causes and Consequences of the Current Evaluation Regime in (Academic) Science,” *Social Science Information* (4)61 (2023), online first, https://doi.org/10.1177/05390184231151610.

Hansson, Nils. *Wie man keinen Nobelpreis gewinnt – Die verkannten Genies der Medizingeschichte*. München: Gräfe & Unzer, 2023.

Hansson, Nils and Thomas Schlich. “Beyond the Nobel Prize: Scientific Recognition and Awards in North America since 1900,” *Notes & Records: The Royal Society Journal of the History of Science* (2024), online first, https://doi.org/10.1098/rsnr.2022.0015.

Ma, Yifang and Brian Uzzi. “Scientific Prize Network Predicts who Pushes the Boundaries of Science,” *Proceedings of the National Academy of Sciences* (2018) 115(50): 12608–12615.

Meho, Lokman I. “The Gender Gap in Highly Prestigious International Research Awards, 2001–2020,” *Quantitative Science Studies* 2 (2021), 976–989.

Münch, Richard. *Die akademische Elite*. Frankfurt: Suhrkamp, 2007.

Nordqvist, Katarina and Pauline Mattsson. “Nobel Prize Awarded Discoveries and Commercialization: The Role of the Laureates.” In *Attributing Excellence in Medicine: The History of the Nobel Prize*, edited by Nils Hansson et al., 188–206. Leiden: Brill, 2019.

Paul, Herman. “What is a Scholarly Persona? Ten Theses on Virtues, Skills, and Desires,” *History and Theory* 53 (2014), 348–371.

Paul, Herman and Jeroen van Dongen, eds. *Epistemic Virtues in the Humanities and the Sciences*. Heidelberg: Springer, 2017.

Sapiro, Gisele. “The Symbolic Economy of the Nobel Prize: How it Counters or Reproduces Modes of Dominations,” *Poetics* (2023), 101. https://shs.hal.science/halshs-04386733/document (accessed June 25, 2024).

Sneis, Jørgen and Carlos Spoerhase, eds. “The Consecrating Power of the Nobel Prize in the Global Literary Field,” *Poetics* 101 (2023), 101821.

Zuckerman Harriet. *Scientific Elite: Nobel Laureates in the United States*. New Brunswick: Transaction Publishers, 1995.

CHAPTER TWO

# Everybody's Searching for a Hero: Controversial Nobel Laureates and the Status of the Nobel Prize

*Gustav Källstrand*

**Abstract**

Since the Nobel Prize has become a ubiquitous legitimizing symbol for science—and the laureates consequently embodiments of the virtues ascribed to science—the impact of a Nobel laureate's socially inappropriate behavior becomes magnified. This can also put the judgement of the prize awarders into question and thus taint the symbolic value of the prize.

Throughout the 20th century, the high status of the Nobel Prize has prevented controversies doing harm to the prize. This paper will argue, however, that this is changing. The status depended to a high degree on the cultural status of celebrating "great men" for their discoveries—and this status has come increasingly under scrutiny during the last decades. Thus, rather than individual decisions or laureates being seen as inappropriate, it is the underrepresentation of women that can undermine the prize's status.

**Keywords:** Nobel Prize, scientific celebrity, prize studies, public understanding of science

Scientific ideas and discoveries are more relatable if they are associated with individual discoverers. We see this in the personalization of scientific theories, like Darwinian evolution, Newtonian mechanics, or Mendelian genetics. Facts and reason are important when we gain knowledge, but emotions and identity also play significant roles for how we perceive the world. We are more likely to accept those with the necessary cultural attributes of credibility as authorities. In science, there is a long tradition of giving credence to the authoritative figure of brilliant individual scientists – the scientific heroes. Over time, they have been inventors, adventurers, enlightened geniuses, inspired dreamers, entrepreneurs, absentminded thinkers, etc. These individuals have been presented to the public in various cultural forms and media, creating idealized images of what a scientist is and ought to be.

The heroism that permeates the public image of scientists has a conduit in the Nobel prize. First awarded in 1901, the prize has become a ubiquitous symbol of excellence and individual brilliance. It honours discoveries, and since the prize has the power to create a link between scientists and their discoveries it also contributes to the personalization of scientific ideas. This process also turns the winners into examples of a type of scientific celebrity: the Nobel laureate.

The Nobel prize has become an institution with the power to routinely create scientific heroes. This power is derived from its status, which depends partly on the credibility of the institution making these choices. Without the acceptance of the scientific community, the prize would lose its legitimacy. But the status also depends on how the Nobel laureates are perceived as living up to their assigned role. Without public acceptance of the laureates as scientific heroes, the prize would risk losing its relevance. This paper offers a discussion on the legitimacy and relevance of the prize based on the effect controversial choices and controversial Nobel laureates has on its status.

The broader implications of this discussion is that the Nobel prize is part of a culture in which the public discourse regarding the legitimacy of scientific knowledge is shaped by the perception of the scientist as a public character, a *persona*.[1] While trust in science means having faith in the methods and practices used to construct the body of scientific knowledge, scientific authority is also linked not only to the expertise but also to the charisma of the individual expert.[2]

## The Nobel Prize and scientific heroes

The Nobel Prize was founded by the inventor and industrialist Alfred Nobel, who wrote in his will that his fortune should be used to reward those who "had conferred the greatest benefit to humankind" through inventions and scientific discoveries, idealistic literature and working politically for peace and disarmament. These ideas were firmly rooted in the same optimistic belief in progress that was expressed at the world's fairs – a belief that reason and knowledge would create both material and moral advancement.

While it was awarded by, and given to, academic scientists – the popular image of the Nobel laureates was influenced by the popular view of the scientist as a heroic individual. The construction of a modern scientific *persona* can be traced back to at least the 17th century, when a conscious effort was made by men of science to create a new social role, separate from alchemists and magicians.[3] These attempts to create a new image is clearly seen in scientific portraits, where men of science were portrayed as solitary thinkers, either separated from crowds or placed in lonely environments such as a cave, a tower, or a study. Painters used the convention of the head-on-hand pose to show them as thinkers who pursued knowledge for its own sake. However, while effective for creating an image of the learned man as a fascinating individual it also, in the words of Ludmila Jordanova, "undermined the claim of men of science to be at the forefront of social change that would benefit humanity as a whole."[4] This otherworldliness could also be perceived as subversive, threatening or laughable. Fictional scientists could be absentminded crackpots, but they could also be the equivalent of dark wizards in fairy tales – in possession of unnatural knowledge and power.[5]

The idea of the scientific genius opened a new path to scientific heroism. Prior to the 19th century, the term *genius* described a trait that could be possessed, but the Romantics made the term refer to a divinely inspired individual. One of the first scientific geniuses created in this fashion was Isaac Newton. During the 18th century, he had been increasingly used by the *philosophes* as a beacon of enlightened rationality, but in the 19th he was portrayed as an almost supernatural thinker, most notably perhaps in William Blake's picture of him as a classical hero.[6] Still, in the Romantic sense of the word, the very idea of a *scientific* genius was a contradiction in terms. Science is objectivity and exactitude, whereas genius is inspiration and passion. From around 1900 and onward, however, the qualities of genius have become the object of measurements, testing, and quantification.[7] Today, intelligence and creativity can be explained both as inherent – and inheritable – talent, and as skills that can be learned through practice, hard work, and determination.[8]

Still, *genius* is not a natural but a cultural category. Geniuses are constructed by their contemporaries – and by later generations – and their meaning and status comes from the place they can play in broader cultural narratives. The same is true for heroes, and, as Ulrich Bröckling notes, the way they are "conceptualized provide information about the meanings, normative orders, and historical images of the social fabric that brought them forth."[9] This means that the way in which historical scientific heroes are constructed affects the scientific persona today. For example, Patricia Fara notes: "As Western society has become increasingly secularized, intellectual ability has gradually displaced saintly dedication as an attribute of greatness, while medical and technological experts have assumed the mantle of miracle workers."[10]

Since the 19th century, an idea of science as objective – that is, disinterested, curiosity-driven, and free from value-judgements – has dominated public discourse. This idea was also closely tied to the practitioners of science; scientists embodied the abstract virtue of objectivity, and their personal integrity thus ensured the integrity of science.[11] In that respect scientists were seen as neutral observers – but at the same time they should also use their knowledge for the benefit of society, and thus live up to the virtues of other middle-class professionals of the time. Science should be both useful and disinterested and the scientists both objective and active participants in public life. Scientists were expected to leave their ivory towers and take part in public life.[12] This continued throughout the 20th century, and neutrality was increasingly seen not as splendid isolation but rather as complacency, especially after World War II.[13]

Since the 19th century, the image of the disinterested scientist was already partly altered by the emergence of engineers and inventors.[14] The trope of the scientist as hero could draw on the image of inventors as benefactors and discoverers as adventurers. There was a co-production of individual scientific and heroic virtues that downplayed the role of collaboration, context, and processes.[15] Scientific discoveries were made by men of action and came though brilliant insights – which could be identified and result in Nobel prizes.

This view was not unopposed. A debate emerged between those who saw science as driven by individuals and inspiration and those who viewed it as a collaborative process. Thorstein Veblen, for example, argued that science should be seen as a collective enterprise in which individual traits were not – and should not be – important, and J.D. Bernal argued that scientific knowledge had the ability to change society because it was shaped by the context in which it was created, and thus answered a need. While voices like this existed, however, the importance of individuals for scientific progress was still stressed by both scientists and the public.[16]

This view was tied to a growing cult of the individual which can be seen from the early 19th century, in the Romantic ideal of the divinely inspired genius, to the late 19th century, in Nietzsche's ideal of the heroic superman. The historian Thomas Carlyle wrote that the genius held the same place in modern society as the prophets had in antiquity.[17] Reaching large audiences, Samuel Smiles' books about great men as role models and sources of inspiration also became influential. He portrayed a new kind of modern hero, not the kings or generals but the striving men of business and industry that built the new industrial world and drove progress. That scientists were included in this group, Steven Shapin notes, was a sign that they were no longer the contemplative thinkers set apart from society, but rather men of action – embodying the same virtues as the other heroes of the middle-class.[18]

Shapin draws on Max Weber's description of science as a vocation, and while Weber downplayed the aspect of the individual as a genius, he also used the concept of charisma to understand how individuals gained influence in social domains. For Weber, charisma was not inherent to the individual – making them predisposed to greatness – but rather constructed by relationships and actions in a social context.[19] How the role of the Nobel laureate was created was an expression of this. The media image of the scientists who won the prize drew on the popular image of scientists as heroes, explorers, and inventors. While the scientific work of the laureates could be described as academic and even boring, their personalities were described as spectacular – making them charismatic, in the sense that they could exert influence.[20]

Social tension could arise from the connection between science and practical benefit. Both academic scientists and engineers were emerging as professional groups which had reasons for maintaining a distance from the other. Scientists saw their work as disinterested and elevated, not to be confused with the practical engineers who chased economic gains, and engineers did not want technology to be seen as merely implementations of scientific ideas, not least since this would make it hard for them to patent processes and inventions.[21] The Nobel prize became a tool for academic scientists in this boundary work, since it awarded basic science and relegated technology and clinical use to the status of applications.[22]

The influence of the Nobel prize can at least partly be explained by its usefulness as a symbol in a culture permeated by an emphasis on the importance of individual scientists' achievements. Adding to this is how during the 20th century the scientific public persona became increasingly celebrified and mediatized. Declan Fahy describes scientific celebrities as public intellectuals who base their influence on knowledge in a specialized field but can expand this authority into other areas by getting access to broader media channels and expressing their views in ways that attract attention.[23]

The contemporary celebrity culture surrounding scientists and entrepreneurs has similarities with the hero-worship surrounding scientists at the turn of the 20th century. Then, there were people like Thomas Edison, Nikola Tesla, Guglielmo Marconi – and Alfred Nobel – who were not considered scientists by the scientific community, but nevertheless were perceived as scientists, or as having a strong connection to science, by the public. Similarly, entrepreneurs today can emphasize their link to science to gain credibility. Their heroism is also similar to that of scientists through its emphasis on creativity, the ability to create ideas and values through imagination and inspiration.[24]

That what are at root romantic ideas of genius and inspiration is present in discourses on progress, technology and science is somewhat paradoxical. Heroism breaks with the ethos of rationalism that permeates modernity, and as Bröckling notes, this makes it an atavism in modern culture. However, this does not mean that it does not exist.[25] Instead, the charisma of the hero can be strengthened by this very break, which explains the durability of the scientific hero, who embodies the contradiction between the systematic scepticism of science and the romantic inspiration of a genius.

Cultural researcher Mattis Karlsson has shown how presenting the geneticist Svante Pääbo as a *discoverer* was a central part of the narrative of the discovery of the Denisovan fossil – reaffirming how important the heroic individual is for the presentation of science in public, even before Pääbo received the Nobel prize in 2022.[26] In this context, the Nobel prize becomes the symbolic affirmation of his status as discoverer – the hero that embodies the search for ancient DNA.

Viewed like this, the Nobel prize becomes an institution that can embody this tension between modernity and tradition. While in 1901, the prize was an invented tradition which used historical forms to legitimize a view of science as the driving force behind progress, in the 2020s it has become the historical form that others can draw on for status.

## Credible heroes

The Nobel prize has the authority to, as it were, appoint scientific heroes. Having given the prize to renowned scientists in the past gives the prize awarders credibility in the present. In that way, it can be both traditional and relevant. This also means that the status of the prize depends on the Nobel laureates in two ways: how well the prize awarders choose the heroes, and how well the laureates perform publicly as heroes. We will examine each of these questions in turn to see if criticism and controversy affect the status of the prize.

### *The prize awarders*

While Nobel prizes in science are often accepted as fair, there are also instances where the choices are questioned, both in the scientific community and, to a lesser extent, in the public sphere. Criticism falls along two main lines. The first is that someone who received the prize did not deserve it, or that someone who deserved the prize did not receive it. The second are cases when a discovery did not deserve a prize, either because it was not important enough or because it had problematic applications. Sometimes these controversies emerge at the time of the award, but they can also surface later.

Criticism directs attention at an aspect of the prize that is usually invisible – the prize awarders and the process of selecting the winners. It is not the prize as an institution that is called into question, since the assumption behind the claim that the "wrong" person or discovery won is that there is a "right" winner. Instead, what is at stake is whether the stewards of the hero-appointing process, the prize awarding institutions, are living up to the status they have been appointed to confer on the laureates.

Rather than undermining the authority of the committees, criticism often seems to reinforce their independence and authority. This is particularly interesting when the prize affects ongoing priority disputes – like the 2008 medicine prize to Luc Montagnier and François Barré-Sinoussi for the discovery of the human immunodeficiency virus. Work on the retrovirus, which would become known as HIV, was started independently in the early 1980s by (amongst others) Montagnier and Barré-Sinoussi at the Pasteur Institute in Paris and by Robert Gallo at the National Institutes of Health in Washington. The two teams competed, collaborated, and fought in public for several years about priority in a conflict that was not only about credit but also about royalties from patents. It also involved national prestige, which resulted in an agreement sealed with a handshake between presidents Ronald Reagan and Jacques Chirac that the two nations would share the royalties and that Gallo and Montagnier should share the credit for the discovery.[27] This, unsurprisingly, did not put the matter to rest. When the Nobel prize was announced and Gallo's name was left out, this was seen as the committee taking a stand in the, then still ongoing, debate. Similarly, when the 2020 chemistry prize awarded Jennifer Doudna and Emanuelle Charpentier the prize for their work on the gene-editing tool CRISPR/Cas9, they were embroiled in priority discussions and patent battles with other teams.[28]

The deliberations in the Nobel committees are kept secret for fifty years which makes determining the extent to which they are concerned with criticism. However, many controversies date further back in time. One of the first historical prizes studied when the archives were first opened for research in the 1970s was the 1922 medicine prize to Frederick Banting and John Macleod for the discovery of insulin. This decision sticks out since it was questioned already at the time by one of the laureates himself. It was Banting who publicly expressed the view that Macleod had not been part of the discovery, and that the prize should have been given to Banting's assistant Charles Best instead. Patents also played a role in this prize not least because Macleod's name did not appear on the patent for insulin, which was issued to Banting, Best and a fourth collaborator, Bert Collip.[29]

The historical record tells a more nuanced story. Macleod was the head of the department where Banting worked, and without his support Banting's work would not have been possible. And while he was not present for the first pivotal experiments, Macleod played an active role in the research. That his name did not appear on the patent was not surprising, Banting's did not at first either (although it was later added for legal reasons). They were both medical doctors and as such not supposed to patent.[30] Archival research also shows that the committee had gained critical knowledge from the Danish Nobel laureate August Krogh, who visited the laboratory in Toronto and stressed the importance of Macleod's leadership.[31]

Despite this work to nuance the perception of the 1922 prize as a "bad" decision, the view lingers in popular accounts. Another case where the committee has been questioned in retrospect is the decision to not include Lise Meitner in the 1945 award of the chemistry prize to Otto Hahn for the discovery of nuclear fission in 1938. They worked as a team with Hahn as the experimenter and Meitner as the theoretician, and she had been crucial for interpreting Hahn's results. That she was not present in Berlin during the work could hardly be blamed on her – she had fled Germany (with the assistance of Hahn) and was in exile in Sweden. Her exclusion from the prize has been interpreted as being the result of opposition from leading members of the Swedish scientific community, especially the physicist Manne Siegbahn who employed her but did not appreciate her work.[32] Again, the archives tell a more complex story. While Siegbahn may not have been friendly with Meitner, he did secure funding for her work and nominated her for membership in the Royal Swedish Academy of Science – which she gained in 1945. But what changes the image even more is that it seems as both the chemistry and the physics committees wanted to award this momentous discovery, and the plan was to award Meitner the physics prize and Hahn the chemistry prize in 1946. But during the vote, the Academy went against these recommendations and gave Hahn the prize – and having already postponed Meitner's award the opportunity was lost, since it would be difficult to award her the physics prize in a different year for the same discovery.[33]

But again, these nuances do not become part of the popular view of the prize. The same pattern can be seen in other controversial or "bad" prizes, like the prize awarded to Paul Muller for DDT, to Egas Moniz, who won the award for use of lobotomy as a psychiatric treatment and Fritz Haber, who was awarded the prize for the synthesis of ammonia but was also the inventor of poison gas during World War I. Further sins of omission include Rosalyn Franklin, not included in the prize for the discovery of the structure of DNA even though her work was used, and Jocelyn Bell, who made the observations which led to the discovery of pulsars for which her supervisor Anthony Hewish won the prize. All these examples can be investigated in greater detail to better understand the difficulties that the committees faced.

However, as mentioned above, these explanations do not usually enter the public narrative, and if they do, it takes a long time and the result is not that the story is changed, but that the controversy is forgotten, since it is no longer a controversy and therefore no longer stands out. Because the key to the interest in the "bad" prizes is that they are the *exceptions*, interesting because a "bad Nobel prize" is seen as an oxymoron. Stories about bad prizes are not used to undermine the legitimacy of the prize awarders – on the contrary, it is precisely the legitimacy of the prize that makes them interesting. Public controversies surrounding a prize has little to do with whether a decision was taken on the "right" grounds or not – it has to do with the public perception of the prize.

That the decision-making process behind the Nobel prize is closed means that it can only be evaluated based on the outcome, not on the methods used to reach the decision. For fifty years, the Nobel committee process is a *black box* which produces Nobel laureates. Bruno Latour described blackboxing as:

> the way scientific and technical work is made invisible by its own success. When a machine runs efficiently, when a matter of fact is settled, one need focus only on its inputs and outputs and not on its internal complexity. Thus, paradoxically, the more science and technology succeed, the more opaque and obscure they become.[34]

Historians and sociologists studying the prize are interested in opening the black box, but to understand the *image* of the prize it is not necessary since the public image of the prize depends on the outcome, not the process. This means that the public's lack of interest in what goes on behind the scenes is a sign that the status of the prize awarders is high. Calls for for the black box to be opened is a sign that the outcomes are no longer accepted – and these calls are usually not raised – not even in cases where the outcome is perceived as controversial.

One way to explain why criticism against the outcome of the black box can still be seen as reaffirming the legitimacy of the boxed process is that, as James English notes, prizes can be seen as a circular system of prestige where criticism plays a function as a marker of relevance.[35] A prize with a high status is *useful* not only for the prize awarders and winners, but also for a broader set of what Sven Widmalm calls "constituencies."[36] The Nobel prize was seen from the outset by its constituents within both the academy and the media as a resource which could enhance discussions on science (and literature and peace) through devices like controversy and celebrity.[37]

*The Nobel laureates*

Scientific celebrities attain their ability to exert influence in fields outside their own expertise – their charisma – through active participation in media and public debate. But the Nobel laureates have their celebrity status bestowed upon them by the Nobel committees, which makes them a special kind of visible scientists. Rather than public figures of their own making, their position is derived from the collective *persona* of the "Nobel laureate."[38] This connection means that the prize plays into how their statements and actions are interpreted and received in public.

We can see this play out in the case of James Watson, who shared the Nobel prize in 1962 with Francis Crick and Maurice Wilkins for the discovery of the structure of DNA. In 2007 he published the latest in a series of autobiographical books, which took the form of life-lessons written as advice to young scientists. In case any of his readers should ever win the Nobel prize, he provides words of caution:

> The moment your prize is announced, you are seen as fair game for petitioners of worthy causes in need of well-known signatories. In lending your name to such appeals, you often find yourself outside your expertise and expressing an opinion no more meaningful than, say, that of the average accountant. You trivialize your Nobel Prize and make future uses of your name less effective.[39]

For Watson, receiving the prize clearly meant acquiring a kind of capital. It is also clear from earlier writings that he saw that this status came from becoming part of a cultural narrative. In a previous memoir from 2001, he described his research as an "adventure story of the best kind,"[40] and displayed himself as an explorer, a heroic scientist. Although Watson was a highly successful and well-established scientist at the time, he still wanted to portray himself as a rebel, an irreverent and unique character rather than a stately man of science. He explicitly set out to tell the story of his life warts-and-all, even though this might upset "readers who want to see me as I am today, rather than the inexperienced and more self-centered person I once was."[41] While Watson clearly thought that this would set him apart, it does follow a pattern of describing Nobel prize winners as unique and atypical scientists which has been familiar in the construction of the Nobel laureate persona since the early 20th century.

Watson had explored this way of presenting himself before, in his 1968 best-seller *The Double Helix* where he also told an "adventure story." He described the discovery of the double helix as a race against other researchers in which he and co-discoverer Francis Crick did not hesitate to get information from other labs through almost any means necessary. The view of science usually put forward in popular science was still dominated by an ethos of describing science as morally uplifting and driven, in the words of Peter Bowler, not "by a desire to build more efficient machines, but by sheer curiosity – although it might, occasionally and by accident, have useful applications."[42] Describing science as a race was a break with tradition, as was his characterization of scientists. "One could not," he wrote, "be a successful scientist without realizing that, in contrast to the popular conception supported by newspapers and mothers of scientists, a goodly number of scientists are not only narrow-minded and dull, but also just stupid."[43] This quote again shows that Watson was aware of how he was using the public image of scientist as a contrast to make himself seem more interesting.

*The Double Helix* was published in 1968, at a time when the public image of science was being renegotiated in the public culture in the wake of the questions raised about the modern view of science as a cumulative and inherently beneficial process by books such as *The Structure of Scientific Revolutions* by Thomas Kuhn and *Silent Spring* by Rachel Carson. This context made the impact of *The Double Helix* more powerful. On the surface, it shared the critical ambition to expose how science worked behind the scenes, inside the black box, and if it had been written by a social scientist, a reporter, or even a junior scientist it might have been read as a swipe against the scientific elite. But instead, it was written by a Nobel prize-winning scientist, who was very much part of this elite – and whose discovery was counted as a paradigm-shifting leap forward. Watson used the questions raised in the public image of science to fashion his own persona as the *enfant terrible* of the biochemical world – an iconoclast who enjoyed being, or being seen as, irreverent and quixotic. He set himself apart from the slightly boring scientists who were doing – to use a term from Kuhn – normal science.

*The Double Helix* was controversial, but Watson's cultural position meant that it did not undermine his status. Instead, it established Watson as an example of the heroic individual prepared to go against established norms to make great discoveries. What Watson did was to use the status of the Nobel prize to make himself seem more like an exceptional scientist. This is the same process as the media used to assign previously unknown researchers the status of Nobel laureates; overstating what set them apart from an expected norm was a way to fast-track the mediatization in which "regular" scientific celebrities engage.[44] Using the status to create a certain amount of controversy can be seen as an important part of how the laureates become and remain interesting public figures. But what Watson wrote in 2007 was that the status from the prize can be both used and misused. There is a point at which controversies are no longer signs of independence but instead of poor judgement – to the detriment of the individual scientist.

Watson eventually caused his own public downfall by making, and repeating, derogatory statements on race and intelligence. The institutions where he worked distanced themselves from him, and he became isolated from the scientific community.[45] Watson is not the only Nobel laureate to go off-board and start promoting offensive or pseudo-scientific views. Physics laureate William Shockley, co-inventor of the transistor, spent the last several decades of his life publicly defending eugenics and, less morally objectionable but nevertheless problematic, chemistry laureate Linus Pauling famously became a champion for vitamin C as a cure for a broad array of diseases, including the common cold.[46] This phenomenon is sometimes called "the Nobel disease," and other examples include ethologist Nikolaas Tinbergen who advocated outdated and potentially harmful therapies for autistic children, Luc Montagnier who has advocated homeopathy, and Kary Mullis who believes in astrology and whose autobiography includes a story of being abducted by aliens.[47]

The heroic scientist is expected to go against the consensus, and stories about scientific celebrities stress their independence – but only up to a point. When they start to express crackpot theories, they usually discredit themselves and therefore lose their charisma. However, the Nobel prize complicates matters. For a scientist like Watson, whose fame goes beyond being a laureate, the Nobel prize can become a problem rather than an asset, since it makes his deviation from the acceptable even more publicized (and criticized). But the status of being a Nobel winner cannot, unlike charisma, be lost. This means that when Ivar Giæver, a scientist who is not well-known, is referred to as a Nobel Prize winner for physics when calling climate change a "new religion," his statement – and those of all the other "climate sceptics" – gains a degree of credibility that it would otherwise not have.[48]

## Legitimacy and cultural narrative

What is the impact of the Nobel prize on science? Historian Svante Lindqvist suggests that to answer this question, historians should study not the laureates, but those who did not win. The premise of the award system "is that there are always more losers than winners," and "the more losers there are... the greater glory for the chosen few."[49] The public naturally follows the winners, but Lindqvist notes that the real impact of prizes on science may well be that those who do not win will look at the prize decisions to think about what kind of work they should do in the future, to put them in the good graces of the Nobel committees.[50] Lindqvist also noted that the criticism levelled against the prize – for example that there are too few fields and that it is unfair that only three people can share the prize – has become routine in the coverage of the prize each year and do not seem to rise to the legitimacy of the prize. One explanation for this is that they are based on misconceptions of the aim of the prize. If the prize aspired to award everyone involved in the greatest discoveries it would indeed be misguided to have limits on the number of recipients – but this is looking at the prize backwards. The prize is given to up to three individuals for a discovery which can be traced to their contributions. That the prize is interpreted as a universal arbiter of credibility is not the fault (as it were) of the prize itself.[51]

While on point, this analysis overlooks that it is precisely the interpretations of the prize that gives it its status. The public – and the scientific community – focus on the winners since this contributes to a narrative of science that gives the media icons and the scientific community legitimacy. In *The Road to Stockholm*, a 2002 book portraying Nobel prize winners, chemist and science writer István Hargittai reflects on the importance of the prize and concludes that: "[t]he Nobel Prize is overrated by the public, but this is not an unwelcome exaggeration. Science needs icons, since it usually suffers from an image of being impersonal."[52] This candid statement captures both the usefulness of the Nobel prize and how the stakeholders are aware of it. The reason stakeholders hold the prize in such high regard is its usefulness – and it would be less useful if controversies could diminish its status.[53]

The Nobel prize does not give an accurate view of how science works. Scholars describe science in terms of process, development, and collaboration, whereas the prize tells a story of events, discoveries, and individual achievements.[54] However, this is not generally seen as a problem by either the prize's stakeholders or its publics since the image of science it projects fits very well in a modern cultural narrative of progress and individualism. Its emphasis on basic science as conferring "benefit to humankind" makes the prize part of a broader effort to legitimize research by convincing the public of the value of free scientific inquiry. The "benefit" of science can come in terms of technological or clinical applications, but for the awarders of the prize, practical benefit has been treated as being a consequence of curiosity-driven science.[55]

Thus, opinions that seek to undermine this narrative is present in the media each year, but more as a news-angle than as serious criticism. One of the most mediatized events surrounding the prize is the Nobel banquet, a media event in which the broadcasters are loyal to the message of the organizers – and the message is, in short, that the Nobel laureates are geniuses.[56]

The prize, and the laureates become part of a strong narrative, and from presenting this narrative in the enduring cultural form of an *invented tradition*. These are "responses to novel situations which take the form of reference to old situations, or which establish their own past by quasi-obligatory repetition,"[57] and the Nobel prize did both. It is, as we have seen, based on a kind of hero-worship that has been part of Western culture since antiquity, and it is also repeated every year in a cycle that makes it seem somehow inevitable.

To damage the status of the prize, a controversy would have to delegitimize one or several parts of this repetitive machinery. However, this is not the role played by the stories of controversial Nobel laureates in the narrative of the prize. On the contrary, they are controversial exactly *because* the status of the prize is so high. When laureates "misbehave," it is they who fail to live up to the ideals embedded in the role they have been ascribed through receiving the prize. The discussion that a controversial prize decision or a misbehaving laureate can generate can be seen as a measurement of the prize's status rather than a threat to it.

That 106 scientists published a letter in *Science* urging that Robert Gallo's contribution to HIV-research should be remembered is a sign that it matters who gets the award.[58] And an inappropriate comment about female scientists by a male professor at a conference would not have been headline news if he had not been a Nobel laureate.[59] If these controversies fail to grab the public's attention, it would be a sign that the prize lacked relevance. Rather than threatening the status of the Nobel prize, controversies are signals of its relevance.

This is not to say that a sufficiently large controversy could not damage the prize, but to do so it would have to make the prize less useful by undermining its place in the greater cultural narrative that we have outlined above. However, the repetitive nature of the prize makes it hard for an isolated scandal to seriously harm its status – the prize gets, as it were, a fresh start each year and the scandals of previous years become anecdotal parts of a history which is overwhelmingly positive.

However, this hinges on the legitimacy that the prize has as part of the narrative which supports it – its power to create credible heroes. This means that there are two ways for it to lose this legitimacy. One is that the process of selecting the laureates is questioned, either through unacceptable choices or on principle. This threat has been present since the prize's inception, and so far, it has not materialized, as we have seen above. The second way is if the people who win the prize are no longer perceived as heroes. Even if the Nobel committees make credible choices, the status of the prize depends on its part in a broader cultural narrative of science progressing through the work of heroic individuals, and to maintain that part the heroes need to be acceptable in a modern media context. The laureates need to be both scientifically and culturally acceptable.

In 1901, awarding "great men" for their achievements was more that acceptable, but one criticism that not only persists but is also increasing the problem of poor gender and ethnic representation. In a public culture where colonial and patriarchal structures were predominant, this was not a problem – on the contrary – but increasing awareness of these issues in both media and the scientific community makes the distribution of the prizes more problematic. And there are no stakeholders who benefit from rhetorically maintaining the idea that science ought to be predominantly white and male. On the contrary, the tendency is to emphasize internationalism and inclusion.

The prize has been described as a "heritage brand,"[60] drawing its credibility from a its (invented) traditional forms. But when the tradition includes increasingly obsolete ideas of which individuals can be seen as legitimate heroes, this history can change from asset to liability. If the prize becomes perceived as traditional in this pejorative way, it becomes less useful. It will no longer be telling the story that the stakeholders need – and a less useful prize becomes less relevant.

Thus, rather than being threats, controversies can serve as the canary in the Nobel coal-mine. As long as the prize decisions are discussed, and as long as a laureate who denies climate change or promotes homeopathy can become a story in the news, the prize is still relevant. The threat would be if a laureate would say something outrageous and no one reacted. This would mean that they no longer fail to live up to the role of the laureate, because there is nothing to live up to. And when heroes can no longer fail us, they are no longer heroes.

## Notes

1 Lorraine Daston and H. Otto Sibum, "Scientific Personae and Their Histories," *Science in Context* 16 (2003), 1–8.

2 Harry Collins and Trevor Pinch, *The Golem: What Everyone Should Know About Science* (Cambridge: Cambridge University Press, 1993), 18.

3 Lawrence Principe, "Alchemy Restored," *Isis* 102, No. 2 (June 2011), 305–312.

4 Ludmila Jordanova, *Defining Features: Scientific and Medical Portraits 1660-2000* (London: Reaktion Books, 2000), 42.

5 Roslynn D. Haynes, *From Faust to Strangelove: Representations of the Scientist in Western Literature* (Baltimore: Johns Hopkins University Press, 1994), passim.

6 Patricia Fara, *Newton: The Making of Genius* (London: Macmillan, 2002), 13–15, 167–181.

7 Darrin M. MacMahon, *Divine Fury: A History of Genius* (New York: Basic Books, 2013), 173–180.

8 Scott Barry Kaufman, *Ungifted: Intelligence Redefined* (New York: Basic Books, 2013), 7–17.

9 Ulrich Bröckling, "Heroism and Modernity," Albert-Ludwigs-Universität Freiburg (2019), 127.

10 Fara, *Newton*, 17.

11 Lorraine Daston and Peter Galison, *Objectivity* (New York: Zone Books, 2007), 191–252.

12 Steven Shapin, "The Ivory Tower: The History of a Figure of Speech and its Cultural Uses," *British Journal for the History of Science* 45, No. 1 (March 2012), 19–22.

13 Cathryn Carson, "Objectivity and the Scientist: Heisenberg Rethinks," *Science in Context* 16 No. 1/2 (2003), 243–269.

14 Jordanova, *Defining Features*, 98, 105.

15 Anders Ekström, "Vetenskapen, medierna, publikerna," in *Den mediala vetenskapen*, edited by Anders Ekström (Nya Doxa; Nora 2004), 15.

16 Fara, *Newton*, 220–221; Steven Shapin, *The Scientific Life: A Moral History of a Late Modern Vocation* (Chicago: The University of Chicago Press, 2008), 1, 6–13, 44.

17 MacMahon, *Divine Fury*, 151, 177, 193, 196.

18 Shapin, *The Scientific Life*, 34.

19 Paul Josse, "Becoming a God: Max Weber and the Social Construction of Charisma," *Journal of Classical Sociology* 13, no. 3 (2014), 270–272.

20 Gustav Källstrand, "Warburg's Dogs: Nobel Laureates and Scientific Celebrity," *Celebrity Studies* 13, no. 1 (2022), 56–72.

21 Paul Lucier, "Court and Controversy: Patenting Science in the Nineteenth Century," *The British Journal for the History of Science* 29, No. 2 (June 1996), 142.

22 Gustav Källstrand, "Science by Nobel Committee: Decision Making and Norms of Scientific Practice in the Early Physics and Chemistry Prizes," *The British Journal for the History of Science* 55, no. 2 (June 2022), 187–205.

23 Declan Fahy, *The New Celebrity Scientists: Out of the Lab and Into the Limelight* (Lanham: Roman & Littlesfield, 2015), 12–13.

24 Cf. Ulrich Bröckling, *The Entrepreneurial Self: Fabricating a New Kind of Subject* (London: Sage, 2016), 102–120.

25 Bröckling, "Heroism and Modernity," 140.

26 Mattis Karlsson, *From Fossil to Fact: The Denisova Discovery as Science in Action* (Linköping: Linköping University, 2022), passim.

27 David France, *How to Survive a Plague: The Inside Story of How Citizens and Science Tamed AIDS* (London: Picador, 2016), 101–104, 137–142, 169–173, 215–216, 225–227.

28 Walter Isaacson, *The Code Breaker: Jennifer Doudna, Gene Editing, and the Future of the Human Race* (New York: Simon & Shuster, 2021), 224–227, 233–240.

29 Michael Bliss, *The Discovery of Insulin: 25th Anniversary Edition* (University of Chicago Press; Chicago 2007), 228–232.

30 Maurice Cassier and Christiane Sinding, "Patenting in the Public Interest: Administration of Insulin Patents by the University of Toronto," *History and Technology* 24, no. 2, (2008), 153–171.

31 Bliss, *The Discovery of Insulin*, 226–229; Erling Norrby, *Nobel Prizes and Life Sciences* (Singapore: World Scientific, 2010), 156–157, 161–164.

32 Ruth Lewin Sime, *Lise Meitner: A Life in Physics* (Berkeley: University of California Press, 1996), 326–329.

33 Robert Marc Friedman, *The Politics of Excellence: Behind the Nobel Prize in Science* (New York: W.H. Freeman, 2001), 244–247; Maria Gunther, "Forskarna som inte fick Nobelpriset," *Dagens Nyheter*, (December 10, 2021), https://www.dn.se/vetenskap/forskarna-som-inte-fick-nobelpriset/.

34 Bruno Latour, *Pandora's Hope: Essays on the Reality of Science Studies* (Cambridge, MA: Harvard University Press, 1999), 304.

35 James English, *The Economy of Prestige: Prizes, Awards, and the Circulation of Cultural Value* (Cambridge, MA: Harvard University Press, 2005), 26.

36 Sven Widmalm, "The Nobel Science Prizes and Their Constituencies," *Public Understanding of Science* 27, no. 4 (2018), 397.

37 Källstrand, "Warburg's Dogs," 3–4.

38 Källstrand, "Warburg's Dogs," passim.

39 James D Watson, *Avoid Boring People: Lessons from a Life in Science* (Oxford: Oxford University Press, 2007), 192.

40 James D. Watson, *Genes, Girls, and Gamow* (Oxford: Oxford University Press, 2001), xxiii.

41 Watson, *Genes, Girls, and Gamow*, xxiv.

42 Peter Bowler, *Science for All: The Popularization of Science in Early Twentieth-Century Britain* (Chicago: University of Chicago Press, 2009), 21.

43 James D. Watson, *The Double Helix: A Personal Account of the Discovery of the Structure of DNA* (New York: W. W. Norton & Company, 1980), 13.

44 Källstrand, "Warburg's Dogs," 7, 13–14.

45 Isaacson, *The Code Breaker*, 386–393.

46 Joel N. Shurkin, *Broken Genius: The Rise and Fall of William Shockley, Creator of the Electronic Age* (London: Macmillan, 2006), 210–215, 252–261; Ted Goertzel and Ben Goertzel, *Linus Pauling: A Life in Science and Politics* (New York: Basic Books, 1995), xv–xvi, 197–204, 211–218.

47 Candice Besterfield, Scott O. Lilienfeld, Shawna M. Bowes & Thomas H. Costello, "The Nobel Disease: When Intelligence Fails to Protect Against Irrationality," *Skeptical Inquirer* 44 (3) (2020), https://skepticalinquirer.org/2020/05/the-nobel-disease-when-intelligence-fails-to-protect-against-irrationality/; Kary B. Mullis, *Dancing Naked in the Mind Field* (New York: Pantheon Books, 1998), 89–95, 152.

48 Kimberley A. Strassel, "The Climate Change Climate Change," *Wall Street Journal*, (June 26, 2009), https://www.wsj.com/articles/SB124597505076157449.

49 Svante Lindqvist, "Salon des Refusés," *Nuncius* 34 (2019), 425.

50 Lindqvist, "Salon des Refusés," 424–425.

51 Svante Lindqvist, *Changes in the Technological Landscape: Essays in the History of Science and Technology* (Sagamore Beach: Science History Publications, 2011), 255–270.

52 István Hargittai, *The Road to Stockholm: Nobel Prizes, Science, and Scientists* (Oxford: Oxford University Press, 2002), p. ix.

53 Jacalyn Duffin, "Commemorating Excellence: The Nobel Prize and the Secular Religion of Science," in *Attributing Excellence in Medicine: The History of the Nobel Prize,* eds. Nils Hansson, Thorsten Halling and Heiner Fanerau (Leiden: Brill, 2019), 30–31.

54 Widmalm, "The Nobel Science Prizes and Their Constituents," 399.

55 Gustav Källstrand, "Science by Nobel Committee," 2–5.

56 Hillevi Ganetz, "The Nobel Celebrity-Scientist: Genius and Personality," *Celebrity Studies* 7, No. 2 (2016), 234–248.

57 Eric Hobsbawn, "Introduction: Inventing Traditions," in *The Invention of Tradition*, ed. Eric Hobsbawn and Terence Ranger (Cambridge: Cambridge University Press, 1983), 1.

58 Giovanni Abbadessa, *et al.*, "Unsung Hero Robert C. Gallo," *Science* 323, No. 5911 (2009), 206–207.

59 Rebecca Ratcliffe, "Nobel Scientist Tim Hunt: Female Scientists Cause Trouble for Men in Labs," *The Guardian,* (June 10, 2015), https://www.theguardian.com/uk-news/2015/jun/10/nobel-scientist-tim-hunt-female-scientists-cause-trouble-for-men-in-labs.

60 Mats Urde and Stephen A. Greyser, "The Nobel Prize: The Identity of a Corporate Heritage Brand," *Journal of Product and Brand Management Research* 24, No. 4, (2015), 318–332.

## Bibliography

Abbadessa, Giovanni, *et al.* "Unsung Hero Robert C. Gallo," *Science* 323, no. 5911 (2009): 206–207.

Besterfield, Candice, Scott O. Lilienfeld, Shawna M. Bowes and Thomas H. Costello. "The Nobel Disease: When Intelligence Fails to Protect Against Irrationality," *Skeptical Inquirer* 44 (3) (2020), https://skepticalinquirer.org/2020/05/the-nobel-disease-when-intelligence-fails-to-protect-against-irrationality/.

Bliss, Michael. *The Discovery of Insulin: 25th Anniversary Edition*. Chicago: University of Chicago Press, 2007.

Bowler, Peter. *Science for All: The Popularization of Science in Early Twentieth-Century Britain*. Chicago: University of Chicago Press, 2009.

Bröckling, Ulrich. "Heroism and Modernity," *Helden. Heroes. Héros Special Issue* 5 (2019).

Bröckling, Ulrich. *The Entrepreneurial Self: Fabricating a New Kind of Subject*. London: Sage, 2016.

Carson, Cathryn. "Objectivity and the Scientist: Heisenberg Rethinks," *Science in Context* No. 1/2 (2003): 243–69.

Cassier, Maurice and Christiane Sinding. "Patenting in the Public Interest: Administration of Insulin Patents by the University of Toronto," *History and Technology* 24, no. 2, (2008): 153–71.

Collins, Harry and Trevor Pinch. *The Golem: What Everyone Should Know About Science*. Cambridge: Cambridge University Press, 1993.

Daston, Lorraine and Peter Galison. *Objectivity*. New York: Zone Books, 2007.

Daston, Lorraine and H. Otto Sibum. "Introduction: Scientific Personae and Their Histories," *Science in Context* 16, no. 1/2 (2003): 1–8.

Duffin, Jacalyn. "Commemorating Excellence: The Nobel Prize and the Secular Religion of Science," in *Attributing Excellence in Medicine: The History of the Nobel Prize*, eds. Nils Hansson, Thorsten Halling and Heiner Fangerau, 30–31. Leiden: Brill, 2019.

Ekström, Anders. "Vetenskapen, medierna, publikerna." In *Den mediala vetenskapen*, edited by Anders Ekström. Nora: Nya Doxa, 2004.

English, James. *The Economy of Prestige: Prizes, Awards, and the Circulation of Cultural Value*. Cambridge, MA: Harvard University Press, 2005.

Fahy, Declan. *The New Celebrity Scientists: Out of the Lab and Into the Limelight.* Lanham: Roman & Littlesfield, 2015.

Fara, Patricia. *Newton: The Making of Genius.* London: Macmillan, 2002.

France, David. *How to Survive a Plague: The Inside Story of How Citizens and Science Tamed AIDS.* London: Picador, 2016.

Friedman, Robert M. *The Politics of Excellence: Behind the Nobel Prize in Science.* New York: W.H. Freeman, 2001.

Ganetz, Hillevi. "The Nobel Celebrity-Scientist: Genius and Personality," *Celebrity Studies* 7, No. 2 (2016): 234–248.

Goertzel, Ted and Ben Goertzel. *Linus Pauling: A Life in Science and Politics.* New York: Basic Books, 1995.

Gunther, Maria. "Forskarna som inte fick Nobelpriset," *Dagens Nyheter*, (December 10, 2021), https://www.dn.se/vetenskap/forskarna-som-inte-fick-nobelpriset/.

Hargittai, István. *The Road to Stockholm: Nobel Prizes, Science, and Scientists.* Oxford: Oxford University Press, 2002.

Haynes, Roslynn D. *From Faust to Strangelove: Representations of the Scientist in Western Literature.* Baltimore: Johns Hopkins University Press, 1994.

Hobsbawn, Eric. "Introduction: Inventing Traditions." In *The Invention of Tradition*, edited by Eric Hobsbawn and Terence Ranger, 1–14. Cambridge: Cambridge University Press, 1983.

Isaacson, Walter. *The Code Breaker: Jennifer Doudna, Gene Editing, and the Future of the Human Race.* New York: Simon & Shuster, 2021.

Jordanova, Ludmila. *Defining Features: Scientific and Medical Portraits 1660-2000.* London: Reaktion Books, 2000.

Josse, Paul. "Becoming a God: Max Weber and the Social Construction of Charisma," *Journal of Classical Sociology* 13, no. 3 (2014): 270–72.

Källstrand, Gustav. "Science by Nobel Committee: Decision Making and Norms of Scientific Practice in the Early Physics and Chemistry Prizes," *The British Journal for the History of Science* 55, no. 2 (June 2022): 187–205.

Källstrand, Gustav. "Warburg's Dogs: Nobel Laureates and Scientific Celebrity," *Celebrity Studies* 13, no. 1 (2022): 56–72.

Karlsson, Mattis. *From Fossil to Fact: The Denisova Discovery as Science in Action.* Linköping: Linköping University, 2022.

Kaufman, Scott Barry. *Ungifted: Intelligence Redefined.* New York: Basic Books, 2013.

Latour, Bruno. *Pandora's Hope: Essays on the Reality of Science Studies.* Cambridge, MA: Harvard University Press, 1999.

Lindqvist, Svante. *Changes in the Technological Landscape: Essays in the History of Science and Technology.* Sagamore Beach: Science History Publications, 2011.

Lindqvist, Svante. "Salon des Refusés," *Nuncius* 34 (2019): 421–426.

Lucier, Paul. "Court and Controversy: Patenting Science in the Nineteenth Century," *The British Journal for the History of Science* 29, No. 2 (June 1996): 142.

MacMahon, Darrin M. *Divine Fury: A History of Genius.* New York: Basic Books, 2013.

Mullis, Kary B. *Dancing Naked in the Mind Field.* New York: Pantheon Books, 1998.

Norrby Erling," *Nobel Prizes and Life Sciences.* Singapore: World Scientific, 2010.

Principe, Lawrence. "Alchemy Restored," *Isis* 102, No. 2 (June 2011): 305–12.

Ratcliffe, Rebecca. "Nobel Scientist Tim Hunt: Female Scientists Cause Trouble for Men in Labs," *The Guardian*, (June 10, 2015), https://www.theguardian.com/uk-news/2015/jun/10/nobel-scientist-tim-hunt-female-scientists-cause-trouble-for-men-in-labs.

Shapin, Steven. "The Ivory Tower: The History of a Figure of Speech and its Cultural Uses," *British Journal for the History of Science* 45, No. 1 (March 2012): 19–22.

Shapin, Steven. *The Scientific Life: A Moral History of a Late Modern Vocation*. Chicago: The University of Chicago Press, 2008.

Shurkin, Joel N. *Broken Genius: The Rise and Fall of William Shockley, Creator of the Electronic Age*. London: Macmillan, 2006.

Sime, Ruth Lewin. *Lise Meitner: A Life in Physics*. Berkeley: University of California Press, 1996.

Strassel, Kimberley A. "The Climate Change Climate Change," *Wall Street Journal*, (June 26, 2009), https://www.wsj.com/articles/SB124597505076157449.

Urde, Mats and Stephen A. Greyser. "The Nobel Prize: The Identity of a Corporate Heritage Brand," *Journal of Product and Brand Management Research* 24, No. 4, (2015): 318–32.

Watson, James D. *Avoid Boring People: Lessons from a Life in Science*. Oxford: Oxford University Press, 2007.

Watson, James D. *Genes, Girls, and Gamow*. Oxford: Oxford University Press, 2001.

Watson, James D. *The Double Helix: A Personal Account of the Discovery of the Structure of DNA*. New York: W. W. Norton & Company, 1980.

Widmalm, Sven. "The Nobel Science Prizes and Their Constituencies," *Public Understanding of Science* 27, no. 4 (2018): 397.

CHAPTER THREE

# Nobel Artefacts: Material Heritage of Nobel Prize Laureates in the Netherlands

*Ad Maas & Louise Lagarde*[1]

**Abstract**

This paper investigates the perception of Dutch Dutch Nobel Prize laureates by the way they have been presented in Dutch museums, in particular Rijksmuseum Boerhaave, the Dutch National Museum of the History of Science and Medicine. How did the images of Nobel Prize laureates that emerged from museum-displays change over time, and how did they reflect social ideals and developments?

**Keywords:** Nobel Prize, Hero, Museum-history, Inclusivity

The tenth of July 1908 was the most triumphant day in the career of the Leiden experimental physicist Heike Kamerlingh Onnes (1853-1926). On that day he and his team for the first time managed to liquefy helium gas, which required a record cold of – at most – 4 Kelvin: the Mount Everest among the gases had been conquered. With this achievement, Onnes opened up for scientific research a whole new temperature domain, at extremely low temperatures.

The vacuum flask containing the liquid helium (a "tea-cup" full), and in which, to put it dramatically, a temperature of just a few degrees above absolute zero had prevailed for the first time in the history of the Earth, was soon replaced by another one. Not long after, a whole new apparatus took the place of the first helium-liquefier.[2]

Although the old apparatus was now out of use, it was not disposed of or – what often happens in experimental science – "cannibalized" (i.e. stripped of reusable parts). Both the flask in which the first liquid helium had been collected and the helium liquefier were carefully preserved in the laboratory as a kind of relic – as material witnesses of a memorable achievement.

The staff of Kamerlingh Onnes' *Natuurkundig Laboratorium* (Physics Laboratory) were well aware of the historical significance of their ground-breaking research. Three years after the glorious liquefaction of helium, in 1911, superconductivity was discovered – the sudden disappearance of electrical resistance at low

Figure 1. "Exhibition" in Kamerlingh Onnes' lab in 1922, with the original helium liquefaction hanging from the wall next to the door (Collection Rijksmuseum Boerhaave, inv. nr. P12920).

temperatures. Although initially the physics community was above all puzzled by this remarkable phenomenon, Kamerlingh Onnes' co-workers carefully preserved for posterity the coils in which superconductivity had for the first time been measured in tin and lead. In 1913 Kamerlingh Onnes was awarded the Nobel Prize for the liquefaction of helium.

Later, other artefacts were added to the "historical collection" of the *Natuurkundig Laboratorium*, such as a blackened piece of glass in which another record cold had been reached, as well as an unsightly glass tube in which helium had been solidified for the first time. In 1922 – the 40th anniversary of Kamerlingh Onnes' doctorate – an "exposition" was even organized in his lab presenting all the material witnesses to his achievements (fig. 1). Scientific research equipment had become cultural heritage. And this is where this paper starts.

We aim to investigate what happened to material artefacts connected to Dutch Nobel Prize laureates from the moment they were appreciated as cultural heritage to the present day. In particular, we will zero in on how these "Nobel artefacts" were used to create a certain image of Nobel Prize laureates, and how this changed over time. The main focus will be on the institution that holds the most artefacts of Nobel Prize laureates: Rijksmuseum Boerhaave, the national museum of the history of science and medicine in the Netherlands, which first opened its doors in Leiden in 1931 and has since grown into one of the most prominent history of science museums in the world.

## The start: Tools become heritage

The ambition to preserve the "historical" objects in the laboratory of Kamerlingh Onnes was one of the reasons behind the foundation of the *Nederlandsch Historisch Natuurwetenschappelijk Museum* (Dutch Historical Museum of the Natural Sciences), as Rijksmuseum Boerhaave was initially called. One of the founders and the first director of the museum, August Crommelin (1878-1965), spent his whole career at the Leiden cold laboratory. Crommelin carried out his PhD research under Kamerlingh Onnes, and later became conservator of the *Natuurkundig Laboratorium*, where he was charged with the task to keep the machinery of the "cold factory" running. After Kamerlingh Onnes' retirement, Crommelin became its deputy director.

As conservator, Crommelin was also responsible for the historical collection of eighteenth-century physics instruments lodged in the attic, which had formerly belonged to such illustrious Dutch professors as Willem Jacob 's Gravesande (1688-1742) and Petrus van Musschenbroek (1692-1761). These instruments were still in use when Kamerlingh Onnes was appointed in 1882. He had them moved to the attic to make room for his own research apparatus. Both the historical collection of – mostly – demonstration instruments, and the relics of modern scientific achievements, would become important pillars of the collection in Crommelin's museum.[3]

The foundation of the *Nederlandsch Historisch Natuurwetenschappelijk Museum* fitted in a broader context of an emerging interest in the national history of science. The burgeoning cultural nationalism of the nineteenth century underpinned the process of nation building of the Dutch nation-state. Glorious events and heroes of the national past were revived in statues, literature, paintings and music. Scientists also became involved in the national effort, taking the initiative to erect statues of revered predecessors like Herman Boerhaave (1668-1738), add memorial plaques to buildings where researchers like Antoni van Leeuwenhoek (1632-1723) and Jan Swammerdam (1637-1680) had lived, and organize festivities to celebrate anniversaries of the birth or death of renowned scientists or important discoveries.[4]

By bringing past achievements to the fore, these scientists also pursued another goal. In the second half of the nineteenth century, natural scientists, supported by progressive-liberal elements in society, challenged the still dominant culture that was focused on a classical education. The main instigator of Dutch "scientific nationalism," the liberal zoologist Pieter Harting (1812-1885), argued that the natural sciences were pre-eminently suited to educate and enlighten civilized citizens.[5] To underline the significance of the natural sciences for society, scientists claimed – in the words of Harting – that these "had accomplished more for the amelioration of humankind than the finest painter or bravest warrior."[6]

In the Netherlands, the debate between advocates of scientific thinking and those who supported a classical education did not result in fierce polemics as it did in Germany and England. The champions of the natural sciences in these two countries founded the Deutsches Museum respectively the Science Museum in the first decade of the twentieth century, with the explicit aim of promoting the values of science and technology among their citizens.[7]

The foundation of their Dutch counterpart took a few decades longer. In 1913, the *Vereeniging voor de Geschiedenis der Genees-, Natuur- en Wiskunde* (Society for the History of Medicine, Science and Mathematics) was founded.[8] One of its objectives was the realization of a museum, which, as we have seen, only opened to the public eighteen years later.

In that same year, 1913, Kamerlingh Onnes was already the fifth Dutch scientist to receive the famous telegram from Stockholm. In 1901 Johannes Henricus van 't Hoff (1852-1911) had received the very first Nobel Prize in Chemistry. One year later, the theoretical physicist Hendrik Antoon Lorentz (1853-1928) and the younger experimental physicist Pieter Zeeman (1865-1943) shared the prize for Physics. And in 1910 it was the theoretical physicist Johannes Diderik van der Waals (1837-1923) – on whose work Kamerlingh Onnes' experiments largely rested – who was awarded the prize for Physics. In the 1920s, two laureates in Physiology or Medicine were added to the list: Willem Einthoven (1860-1927) in 1924 and Christiaan Eijkman (1858-1930) in 1929. This was the Nobel state of affairs when the museum opened.[9]

As this "mild shower" of Nobel Prizes shows, the promotion of science by Harting and his contemporaries appears to have paid off. From 1875 the Dutch sciences experienced an unprecedented period of success, with the Nobel Prizes being just the tip of the iceberg (in astronomy, too, for which there exists no Nobel Prize, Dutch scientists were for instance remarkably successful). No longer did the Dutch have to look in the – distant – past to find their scientific heroes.

Indeed, the museum's collection included artefacts from the recent past from the very beginning. Besides the equipment from Kamerlingh Onnes' cold research, it boasted the electromagnet used by Zeeman for the research which won him the Nobel Prize,[10] as well as a few personal items and documents of Lorentz. Some years later, key pieces from the scientific oeuvres of Van 't Hoff and Einthoven were added to the collection. Among the items donated to the museum by the famous botanist Hugo de Vries (1848-1935) was a herbarium which he had specially prepared for the occasion.[11] Other interesting "recent" collection pieces were the remnants of the experiment Wander Johannes de Haas (1878-1960) had conducted with Albert Einstein (1879-1955) (Nobel laureate, not Dutch), leading to the discovery of the Einstein-De Haas Effect. Not only co-workers at the Kamerlingh Onnes Laboratory appreciated the value of preserving the artefacts of successful research and personal items of scientists.

### Nobel artefacts as relics (1931-1949)

The then still modestly sized Dutch museum for the history of science was housed on the upper floor of the former anatomical lab of Leiden's academic hospital. Its rooms were organized thematically. There was a room for astronomy, one for physics and one for physiology, et cetera. In the room for physics, a few low showcases close to the windows and two larger display cabinets contained instruments and recollections of "major Dutch physicists of the latter years: *H.A. Lorentz, H. Kamerlingh Onnes, P. Zeeman* and others" (fig. 2). To be sure, not only Nobel Prize laureates belonged to these "major Dutch physicists": research equipment for remarkable experiments by Willem Keesom (1876-1956), who was one of Kamerlingh Onnes' successors, together with De Haas, were also on display, as well as the artefacts of the Einstein-De Haas experiments.[12] (Actually, a few months after

Figure 2. The helium liquefaction in the initial exhibition (Collection Rijksmuseum Boerhaave)

Figure 3. Museum display in the 1938 (Collection Rijksmuseum Boerhaave)

the opening of the museum, Einstein and De Haas themselves were able to admire the remnants of their own experiments when they visited the museum together).

One photograph demonstrates that some years later, after Einthoven's equipment had been added to the collection in 1939, the arrangement had already been changed. A roll call of fame seems to have been erected along one of the walls of the physics room, with showcases containing the prototype of Einthoven's ECG equipment (which arguably would have been more appropriately placed in the physiology room) and highlights from the Kamerlingh Onnes Laboratory. The portraits on the wall are difficult to make out in the photograph, but seem to include those of Lorentz, Kamerlingh Onnes, Van der Waals, Van 't Hoff and perhaps also Einthoven (fig. 3).

Rather than informing the general audience about the blessings of science and technology, which was the policy of the Deutsches Museum and the Science Museum, anxieties about the preservation of scientific material heritage was the main *raison d'être* of the Leiden museum for the history of science.[13] Both its founders and its board members belonged to a small circle of academics – science and medicine professors, physicians, dentists – who were able to appreciate the historical significance and ingenuity of scientific and medical instruments, documents and preparations. This was the audience which the museum primarily catered to. It thus became a place for a relatively small target group of (as Crommelin called them) “educated people” (*ontwikkeld publiek*), though he also perceived an educational task for the museum and welcomed the eager-to-learn (*leergrage*) youth.[14] Only reluctantly, it seems, he also acknowledged a broader educational goal when he stated in 1947 that “a visit to judiciously arranged museums and exhibitions can help to educate our people towards states of development and civilization, higher than those that can be accomplished by other cultural factors, such as film, radio, jazz, bridge, cross-riddles.”[15]

In an early museum guide, Crommelin formulated the main aim of the museum as the “[p]romotion of the study of the history of science.” But he added to this that the foundation of the museum should also be regarded as an:

> [...] act of piety owed to the memory of many great Dutchmen, who have contributed so much to the fame of the Netherlands through their researches in the field of natural sciences.[16]

History of science, then, was in Crommelin’s perception largely a history of great men (“great Dutchmen”), as it had been earlier in the days of Harting. But at the beginning of the twentieth century, the celebration of scientific heroes of the seventeenth and eighteenth centuries had been replaced by a greater emphasis on more recent scientific celebrities. In 1915, a large statue was erected in Rotterdam for Van ’t Hoff (who had died four years earlier). Van ’t Hoff was portrayed seated, and the statue was placed on a high pedestal, presenting the Rotterdam-born chemist as a lone genius highly elevated above the masses (fig. 4). No less detached from the ordinary world, the great physicist Lorentz was depicted in the monument erected in his native town Arnhem in 1931 as standing solemnly at the end of a flight of stairs (fig. 5).

These depictions resonated with a rather romantic view of scientific key figures current at that time. The view of the scientist as an unapproachable genius is reflected in the exuberant accolades written by the female student E.B. Venema about (and to!) her supervisor, the unassuming and introverted physicist Zeeman:

> We have had the privilege of seeing a great man up close. Not the way one gets to see the queen: it is over in the blink of an eye with only a vague impression of colour remaining in the memory. But really seeing, getting to know, getting to know so well that we know your judgement, your criticism and your admiration regarding many things, that we think in the case of an optical phenomenon: “How wonderful the Professor would find this sight” [...] We have seen up close and therefore sensed the remote distance.[17]

Zeeman himself had several portraits of famous scientists who were a source of inspiration for him on the walls of his office and corridors of his laboratory: “I keep studying their faces: Ampere, Faraday, Hertz. – Here the Dutch: Onnes, Lorentz, Van

Figure 4. Postcard of Van 't Hoff's statue in Rotterdam (Unknown photographer; Collection Rijksmuseum Boerhaave. P10924)

Figure 5. Postcard of Lorentz's statue in Arnhem (Unknown photographer, Collection Rijksmuseum Boerhaave, inv. nr.P18146)

der Waals, Van 't Hoff. – Just look at that energy of spirit blazing from the eyes."[18] Observations like these seem to contradict assertions like the one made by cultural historian Lotte Jensen that Dutch culture is too down-to-earth to nourish national heroes.[19]

At the beginning of the twentieth century, Dutch scientists no longer felt the need to appeal to the history of science to promote the cause of science in society and foster a sense of self-esteem among themselves. The history of science had been more or less relegated to the status of a noncommittal leisure activity, in which also the Dutch Nobel laureates could be the topic. The chemist Ernst Cohen (1869-1944), a member of the advisory board of the Leiden museum, for example wrote an extensive biography on his mentor Van 't Hoff, which has been qualified by Van 't Hoff's recent biographer Rob van den Berg as a Great Man history and "a eulogy of labour and love."[20] In biographical works, great scientists were usually depicted uncritically as outstanding figures who, despite setbacks, disappointments and opposition, managed with immense effort, determination and ingenuity to achieve stunning results[21] – echoing traits of the archetypal literary hero as discussed by Daniella Link in this volume. Thus, Crommelin's description of the museum as an "act of piety" to great Dutch scientists was in line with the – romanticized – adoration of men of science typical of these historical works. Preserving and displaying their artefacts as "relics" – a word Crommelin literally used – was part of this veneration:

> All these objects, relics as we may call them, are on display in the museum and are open to the reverent and admiring gaze of researcher and layperson alike. Neither of them will they fail to impress, and so shall the spirits of our great countrymen continue to speak to us, the admiring posterity.[22]

In conclusion, artefacts of Nobel Prize laureates were an essential part of the Leiden museum for the history of science from the very beginning. They were seen by the museum goers, who mostly belonged to academic, "educated" circles, as "relics" of brilliant heroes of science who were venerated in a way worthy of their elevated status.

## Geniuses for broader audiences (1949-1970)

In 1947, the *Nederlandsch Historisch Natuurwetenschappelijk Museum* became a national museum. From now on it was largely funded by the state, becoming – in principle – an instrument in the state's cultural politics.[23] Its name was changed into *Rijksmuseum voor de Geschiedenis der Natuurwetenschappen* (National Museum for the History of Natural Sciences). Two years later Crommelin retired. He was succeeded by Maria Rooseboom (1908-1978), a biologist – and specialist in the history of microscopy – who had been working in the museum for years and already had a reputation as the "driving force behind the growth of the museum."[24]

Rooseboom took the helm in an exhausted country that was in need of reconstruction after the devastations of the German occupation. Science and technology were rendered an important role in the recovery of the Netherlands, yet at the same time the war had fostered anxieties about the detrimental effects these two had on society. In intellectual circles, such anxieties had already been expressed before the war. Science and technology, it was felt, transformed society at a fast rate, but general culture was not able to keep up with these changes, resulting in the superficiality and spiritual emptiness of the "mass man" (*massamens*), in materialism and in the flattening of culture.[25]

World War II, which had caused mass destruction through modern technology and had been ended by two atomic bombs, added a more broadly conceived, social urgency to the reconciliation of culture and science. Historians of science argued that their field was pre-eminently able to contribute to overcoming the gap between the two intellectual domains. The famous Dutch historian of science Eduard Jan Dijksterhuis (1882-1965) spoke of the historian of science as the "ferryman" who could bridge the shores of the humanities and the sciences.[26] As F.J. Duparc and W.A. van Es report, Rooseboom's museum for the history of science was willing to contribute and help bridge "the gap that had arisen here in this country between professionals in the natural sciences and those in the humanities."[27]

Such efforts should not only be directed at academic audiences. While Crommelin had intended to offer the collection to a more exclusive target group, Rooseboom's principal aim was to open up the museum to "less educated people" as well. As described in the museum's annual report of 1950, Rooseboom considered the main purpose of the museum to be that of an educator and promoter for the growing number of people with an interest in the history of science.[28] She was well aware that the collection of the museum – in her view one of the most important in the world – was hard to popularize,[29] but Rooseboom – known as a hard worker – was ready to take up the challenge.

Museums in the Netherlands after 1945 became more popular and increasingly boasted higher visitor numbers (the number of Dutch citizens visiting at least one museum annually rose from 12% in 1945 to 59% in 1975).[30] Thanks to Rooseboom's dedication, the *Rijksmuseum voor de Geschiedenis der Natuurwetenschappen* kept pace with the trend. For the first time fully-fledged temporary exhibitions were organized, and Rooseboom soon learned "that it is possible to present a topic in such a way that the specialist, as well as the superficially interested finds something to suit them." Schematic drawings, reproductions, texts and demonstrations were added to explain the objects. In Rooseboom's words, the museum changed from a "study collection" (*studieverzameling*) into a "didactic museum" (*didactisch museum*).[31]

That Nobel Prize laureates could play a major role in this "didactic museum" is shown by the fact that the very first full exhibition, in 1953, was about Kamerlingh Onnes and Lorentz. This exhibition, showcasing the work and life of both scientists, was held in honour of the centennial of the birth of the two Leiden physicists.

The names of Kamerlingh Onnes and Lorentz, who had grown into public figures in the interwar period (especially Lorentz), may have served to attract audiences less familiar with liquid helium or electromagnetics. Posters were designed and distributed to promote the exhibition, another first for the museum. Rooseboom contacted newspapers requesting them to pay attention to the exhibition. The involvement of Albert Einstein added extra lustre. At the request of Rooseboom, the world-famous physicist donated the letters he had received from Lorentz and Kamerlingh Onnes to the museum (they are still in the collection).[32] He also wrote a moving portrait of Lorentz, whom he had regarded as a kind of father figure.[33]

The space devoted to Kamerlingh Onnes in the exhibition comprised instruments and machinery, with the original helium liquefactor as pièce de résistance. Lorentz's part of the exhibition space focused more on letters, essays, notes and the like.[34] Scientific objects and papers were complemented by personal items such as photographs of the two scientists in their youth, school diplomas and awards (including the Nobel Prize medals).[35] Particularly timely was the presentation of Lorentz's meticulous calculations for the precise location of the Afsluitdijk, the dam that closed the Zuiderzee, an inlet from the North Sea that was consequently turned into a freshwater lake (the IJsselmeer). Just a few months earlier, on 31 January-1 February 1953, the North Sea coast had been hit by a devastating storm surge, killing almost two thousand people – but indeed the Afsluitdijk had held and the coasts of the IJsselmeer had been spared. The disinterested work of Nobel Prize laureates could even save lives.

The newspaper reviews of the exhibition highlighted in particular the remarkable giftedness, traits and feats of both physicists, achieved through hard labour, vision and perseverance (Onnes) or, even more stunningly, without any apparent effort (Lorentz). One of the objects often referred to in the articles was a table of logarithms used by Lorentz to teach himself solving logarithmic equations at the remarkably young age of ten – a striking testimony to his precocious ingenuity.[36] Both through the exhibition itself and the newspaper coverage, the genius culture of the "grandmasters of Dutch science" or the "two scientists of world fame" was brought to a wider public.[37]

Actually, the year before, in 1952, the centennial of the birth of another Nobel Prize winner, namely Van 't Hoff, had also been commemorated with a small exhibition that was shown in Amsterdam and Delft before it arrived in Leiden for just a fortnight. The display was curated by Rooseboom and Van 't Hoff's former student W.P. Jorissen (1869-1959).[38] Aimed at scientific circles, it was the type of event that could as well have been organized under Crommelin's directorship.

Rooseboom's ambitions for the museum were severely hindered by a lack of space. To host a temporary exhibition, parts of the permanent collection had to be removed from the cramped museum spaces, which even then sometimes proved to be too inconvenient. In 1953 and in 1960 still, two small exhibitions about Einthoven were organized. Following these two, no exhibitions featuring Nobel laureates were held in Leiden for quite some time. In a refurbishment of the permanent exhibition in 1959, a wall was erected to showcase the Netherlands' greatest scientists, namely Boerhaave, Carolus Linnaeus (1707-1778) (actually a Swede), Kamerlingh Onnes, Lorentz and Einthoven. On display in this somewhat arbitrary gallery were objects that would keep their "memory intact": portraits, instruments, books and Nobel Prize medals.[39]

Apart from the donation by Einstein, a few other notable acquisitions were also made, such as the Nobel medal awarded to Einthoven (on semi-permanent loan), a silver medal with the buste of Van der Waals and a large collection of documents and medals formerly owned by Van 't Hoff.[40]

In short, thanks to Rooseboom's dedicated directorship, broader, "less educated" audiences were able to become acquainted with artefacts of the Dutch Nobel Prize laureates, and the genius culture surrounding them, at a time when the history of science saw itself as a cure to counter the sense of alienation resulting from modern, technological and scientific, progress.

## No more heroes? Part 1 (1970-2000)

The perception of Nobel Prize laureates as scientific geniuses did not significantly alter even after World War II. Only in the 1970s and 1980s, a shift must have occurred under the surface, as became clear in 1991 when the museum moved to its current premises in the former Caecilia Guesthouse after a long preparatory period of two decades.

In the new permanent exhibition the objects themselves took centre stage. They were meticulously placed in showcases in a serene, aesthetic setting without any bells and whistles. Any impression of sensationalism was carefully avoided. Tellingly, dazzling but "unscientific" collection items like the pen Einstein used when writing his Theory of General Relativity, or the moonrock samples from the Apollo 11 and 17 missions – imbued with the heroism of the astronauts who conquered the moon – remained in storage (even though the total exhibition space had become five times as large).[41]

The collection of Museum Boerhaave – as the museum was renamed – was presented in the new building in a cultural-historical context (Renaissance, Golden Age, Enlightenment). The work of the Nobel Prize laureates was embedded in the so-called "Second Golden Age" of Dutch Science, as historians had dubbed the successful decades around 1900. This Second Golden Age and its background were detailed in the introductory text to the gallery on chemistry where – to start with – Van 't Hoff was presented as the first of the Nobel Prize laureates. In the subsequent rooms on modern physics (Kamerlingh Onnes, Zeeman), medical science (Einthoven) and microscopy (Zernike), instruments of other Nobel Prize laureates were exhibited.

The detached and intellectualistic style of display did not really lend itself to putting heroes of science on a pedestal. The focus was on scientific instruments, hardly on the persons behind them. A few portraits were shown, but tellingly personal items from the collections of Van 't Hoff, Kamerlingh Onnes and Lorentz were not on display.

The new museum display was developed against the backdrop of a changing society. Starting in the "anti-authoritarian" 1960s and 1970s, egalitarian ideals emerged that were difficult to reconcile with the devotion for past heroes of science as had been entertained by previous generations.[42] A more sober view on past heroes also resonated with contemporary historiography, which in the 1980s and 1990s pursued a rather disinterested and detached approach. It became a "game for intellectuals," with Dutch historians assuming the role of "distanced, mild commentators."[43] None of the temporary exhibitions in this period drew on the Dutch Nobel Prize laureates – except for a modest presentation on the occasion of the centennial of Van 't Hoff's Nobel Prize in 2001. Actually, after ambitious exhibitions on Boerhaave (1968) and Christiaan Huygens (1979), none of the expositions from then on specifically focused on "Dutch" scientists or scientific achievements. Nor were there many important acquisitions in the domain of Nobel Prize laureates in this period (though the Van 't Hoff collection was still extended considerably).

By presenting the collection squarely in a cultural-historical setting, the museum drew on insights that had been developed by Dutch historians of science. Surprisingly enough, the first generations of professional historians of science had largely ignored the Nobel Prize laureates in their work.[44] The laureates were left in the hands of scientists who liked to engage with their discipline's past and their illustrious predecessors. As we mentioned earlier, such amateur historians of science entertained a slightly romantic view of their idols. This tradition persisted after World War II, yielding volumes with grand titles like *Helden der wetenschap* (Heroes of Science) (1946), *Grote Nederlanders bij de opbouw der natuurwetenschappen* (Great Dutchmen who built the Natural Sciences) (1948) and *Pioniers der natuurwetenschappen* (Pioneers of the Natural Sciences) (1983).

Figure 6. Exhibition Quest for Absolute Zero (2008) (Collection Rijksmuseum Boerhaave)

Only in the 1980s did a "professional" interest in the Nobel Prize laureates emerge. It was the sociologist and historian of science Bastiaan Willink who coined the term "Second Golden Age."[45] At the same time Amsterdam-based historian of science Anne J. Kox began a decades-long engagement with (in particular) Lorentz, whereas his Utrecht colleague Harry Snelders studied Van 't Hoff's contributions to chemistry. In his standard work on Dutch history of science *In het voetspoor van Stevin* (1986), historian Klaas van Berkel gave ample attention to the decades around 1900 and the Nobel Prize laureates. In the 1990s, several PhD students were writing their theses based on the archives of persons and institutions related to the Second Golden Age. Scholarly biographies of Nobel Prize laureates also started to appear. In this professional engagement with the Second Golden Age, the adoring stance of the amateur scientist-historian was exchanged for an attitude of professional distance (although, ironically, the grand name "Second Golden Age" might suggest otherwise).

The detached museological approach pursued by the renewed Museum Boerhaave thus coincided with developments in historiography, more specifically the historiography of science. Actually, Van Berkel's *In het voetspoor van Stevin* was used as a kind of guide for the refurbishment of the museum. The Nobel Prize laureates thus entered the new millennium no longer as objects of uncritical devotion, but more neutrally as outstanding exponents of a distinct era in the national history of science.

The 2008 exhibition *Jacht op het absolute nulpunt* (Quest for Absolute Zero) to celebrate the centennial of Kamerlingh Onnes' liquefaction of helium heralded a break with the understated type of exhibition display. Quest for Absolute Zero focused on the, sometimes tense, competition among scientists to reach ever lower temperatures (and thereby liquefying gases), with Kamerlingh Onnes and his successors as the key figures in this "quest." Rather than being serenely placed in showcases, the instruments were embedded in theatrical designs that depicted the different episodes, using dazzling contextual materials like huge photographic blow-ups, film, animations and soundscapes (fig. 6). The exhibition surpassed all previous ones in terms of visitor numbers and publicity, and that was exactly what it aimed to do.

## The brand Nobel (2007-2017)

With the help of an "audio archeologist," Rijksmuseum Boerhaave curator Bart Grob managed to recover in November 2022 the voice of Willem Einthoven, which had been recorded on wax cylinders that were part of the Einthoven collection that had been kept in the storage spaces of Rijksmuseum Boerhaave for decades. Although the recordings did not prove to be extremely insightful, the thin sound of Einthoven's voice sparked an enormous response in the Dutch media. Programmes ranging from the science bulletin *Atlas TV* to news radio programmes, even the *Acht Uur Journaal* (the primetime news programme on national television) presented an item about Einthoven's retrieved speech segment. National newspapers and news websites also devoted attention to the discovery, with headlines like: "Voice of Dutch Nobel Prize Winner Heard after Almost 100 Years."[46] Indeed, besides "voice" there was one other word consistently popping up in the headlines: "Nobel Prize winner." Without Einthoven's status as a Nobel Prize laureate – as colleagues in the communication department of the museum also insist – this result would never have gone "viral" – to use modern terminology – the way it did.

Surprisingly, perhaps, the Nobel Prize's potential for public appeal had only been discovered, or at least purposely utilized, quite recently. In 2011 an exhibition was organized which explicitly drew on the Nobel factor: *Nobelprijs NL* (Nobel Prize NL). It was the first time the Dutch Nobel Prize laureates were presented as such in the museum. A few years later, following a major refurbishment of the museum, a large "Nobel showcase" became a feature of the new permanent exhibition, in which objects and stories of (all) the Dutch Nobel laureates were brought together.

In the final decades in the twentieth century, collection preservation had been considered the museum's main task, both by the museum management and the Ministry it was accountable to. This changed after the turn of the century. Under the influence of neo-liberal ideology and its trust in market forces, museums were increasingly judged by their visitor rates and the income they managed to generate. These were the output targets for alleged good entrepreneurship.

The serene and (in the intellectual sense) somewhat elitist approach of Museum Boerhaave now no longer sufficed if it were to survive in the new political and social reality. It simply did not attract enough attention and visitors. The new director Dirk van Delft (b. 1951), a former science journalist, allayed fears of becoming too lowbrow or sensationalistic. History of science had to be made accessible and also easy to be picked up by the mass media, the importance of which was becoming increasingly evident. One of the ways to do so was utilizing, without reservations, iconic people (such as Einstein, Newton, Darwin, Curie) and phenomena with a wide public appeal in exhibitions and other activities. Whereas many Dutch people may never (or only vaguely) have heard of the Second Golden Age and scientists like Einthoven or Zeeman, virtually everyone is familiar with the Nobel Prize. Why not exploit, in this commercial era, this successful "brand" to the full?

Nor did the museum shy away from introducing (moderately) nationalistic undertones. Besides the Nobel Prize, the title *Nobelprijs NL*, after all, also highlighted the Netherlands (NL). As the exhibition guide of this "Hall of Fame of Dutch Nobel Prize winners" explained, the nineteen Dutch Nobel Prize winners ("not a bad score") revealed "the perseverance and creativity of our most famous scientists of the twentieth and twenty-first centuries."[47]

In the 1990s, debates on national identity and nationalism had flared up in intellectual circles, following a period in which anything regarding the nation-state, nation building and nationalism had been regarded with suspicion (and as outdated). The reappraisal of national identity was prompted by tensions caused by immigration and the multicultural society, secularization and the ongoing European integration. Intellectuals argued for a kind of "civilized nationalism" to counter fragmentation and the disintegration of the social fabric of Dutch society. Traumatic events after the turn of the century: 9/11 and the murders of politician Pim Fortuyn (1948-2002) and filmmaker Theo van Gogh (1957-2004), both of whom were known for their critical stance toward Islamic religion and culture, only reinforced the perceived urgency of the matter. Among more conservative groups a nationalistic attitude emerged which has been coined by Rogier van Reekum as "neo-patriotism."[48]

In the renewed attention to national identity, the inclusion of famous Dutch achievements and persons past and present in the history curriculum or – for that matter – museum displays, could serve to evoke "recognition and response."

Together with soccer players and major water management projects, the Dutch Nobel Prize laureates thus could contribute to a positive identification with Dutchness and Dutch history.[49]

The year *Nobel Prijs NL* was organized, 2011, was an eventful year for the museum. The conservative Rutte-Verhagen administration, endorsed by the extreme right populist party PVV, almost closed the museum down. In the view of the responsible State Secretary, Halbe Zijlstra (b. 1969), it had demonstrated insufficient progress in the amount of income it had been able to generate from visitors and sponsors in the preceding years. Actually, *Nobel Prijs NL* was one of the activities the museum organized to show the value and relevance of its collection in a bid to rescue the museum from closure.

The medium-sized exhibition devoted a showcase to each Nobel Prize laureate and presented one or more objects or images related to the laureate in question. Sometimes these objects were rather anecdotal – such as the building blocks Gerard 't Hooft (b. 1946; shared Nobel Prize in Physics 1998) played with as a child, or a spray can of Paul Crutzen (1933-2021; shared Nobel Prize in Chemistry 1999). The latter was a nod to his research on chemical, atmospheric processes related to the "hole in the atmosphere," which appeared to be caused partly by the propellants in spray cans. Van Delft had requested the Nobel Prize laureates still living to contribute an item (such as the two above). It was an extra bonus if they were afterwards donated to the collection (has happened with Crutzen's spray can).

Similarly, and partly containing the same objects, a large showcase was introduced in the permanent exhibition to display artefacts of the Dutch Nobel prize laureates when the museum reopened after a major refurbishment in 2017. The detached intellectualism of the previous display was no longer to be found in the new exhibition. One of the objectives of the new presentation was to present "more flesh and blood," as the annual report of 2017 succinctly stated. A touchscreen computer provided information – scientific as well as personal – about the Nobel Prize laureates featured in the display, aiming to highlight qualities like "guts, curiosity, creativity and perseverance."[50] It was during the refurbishment process, incidentally, that the museum was given its current name: Rijksmuseum Boerhaave.

### No more heroes? Part 2 (2018-present)

The story of Nobel Prize laureates in Rijksmuseum Boerhaave ends ironically. As we write these words, their artefacts have disappeared altogether from the permanent exhibition. The enigmatic molecule models of Van 't Hoff, the iconic gull's head models of Niko Tinbergen (1907-1988), the silver tray offered to Lorentz to commemorate the fiftieth anniversary of his doctorate – they are currently all kept in the museum's storage space, as is the latest significant "Nobel" acquisition: a collection of artefacts of Peter Debye (1884-1966; Nobel Prize in Chemistry in 1936), which was added to the collection in 2019.

The reason for their (temporary) removal was a refurbishment of the gallery which contained the large Nobel showcase. The topic did not sit well in the new design, while it also proved difficult to find a suitable new place right away.

It is not just a lack of space, however, that complicates their return to the museum galleries. Calls for more diversity and inclusivity in museum displays signify a change in the public debate on identity. Whereas around 2000 it concerned national identity, fifteen years later the bias has changed to group identity. Vocal calls to end the underrepresentation of certain groups like women, people of colour and non-western cultures, challenge the unproblematic replacement of a bunch of white men as paragons of science in the museum display by more of the same. In this perspective, Nobel Prize laureates have transformed from national scientists into "privileged" white men whose representation in the museum only serves to underline the neglect of other, marginalized groups.

Moreover, the narrative of the scientist as isolated genius is being challenged. The scientific practice is increasingly seen as a collaborative effort. As historians of science now argue, have scientists not always been exponents of their times, part of networks of knowledge exchange and building on the work of others rather than solitary actors who developed their brilliant insights on their own, as was the prevailing image so far (see also the contribution of Heeren in this volume)? Scientists themselves also seem to realize this. When Ben Feringa (b. 1951) was awarded the (shared) Nobel Prize in Chemistry in 2016, the photo that was posted on social media showed him surrounded by his whole team, in striking contrast with the lone statues of Van 't Hoff and Lorentz discussed earlier.

How to present Nobel Prize laureates both as outstanding figures and as part of a larger research community? How to present a parade of white men in a more inclusive manner? These are dilemmas that Rijksmuseum Boerhaave, as well as other museums exhibiting artefacts of Nobel Prize laureates, have to face.

## Outlook

In 2018, a special issue of the journal *Public Understanding of Science* was devoted to the media – such as journals, film, television and internet – in the context of public understanding of science. Media coverage of Nobel Prizes disseminates images of science and scientists to the wider public, and thus contributes to the impact of the Nobel Prize. As we claim to have shown in this paper, museums, too, should be regarded as part of what has been termed the "Nobel constituency."[51]

Also, presentations of Nobel collections in museums should be taken into account if we want to understand the use of Nobel Prize images and how these changed over time.

This paper is, as far as we know, the first systematic exploration over a longer period of time of how science museums have deployed artefacts of Nobel Prize laureates. An open question, then, is to what degree the pattern that we have distinguished in the museological representation of Nobel Prize laureates, reflects a more general trend or just concerns a Dutch (or even Leiden) idiosyncrasy. To determine this, we would need to know how "Nobel" artefacts have been used in other countries. The *Ehrensaal* (Room of Honour) opened by the Deutsches Museum in 1925 to celebrate the "magnificent achievements" (*Großtaten*) of the "intellectual heroes" (*Geistesheroen*) in science, engineering and industry, evoked the same kind of romantic reverence as the Dutch museum for the history of science, well until after World War II. It was meant to inspire awe for the "Great Men, whose ideas or whose labour have led to many of the masterpieces on display in the museum's galleries." At the same time, the presentation in the *Ehrensaal* was influenced by specific, turbulent political and social events that put their mark on twentieth-century Germany (World War I, the Weimar Republic, the Nazi era, the division into East and West Germany). Thus, national circumstances impacted on what may be general, international trends in the way Nobel Prize laureates were presented and used in a museum context.[52]

## Dutch Nobel Prize laureates elsewhere

Artefacts of Dutch Nobel Prize laureates are also kept in museums other than Rijksmuseum Boerhaave. The University Museum of Groningen boasts significant objects related to the phase contrast microscope, the invention which earned Frits Zernike a Nobel Prize in 1953, and to Ben Feringa. It put the Nobel Prize front and centre for an exhibition on Frits Zernike in 1988 called *Frits Zernike: Groninger Nobelprijsdrager 1888-1966*, demonstrating how the Nobel status could be a crowd puller.[53] The two Groningen Nobel Prize laureates heavily featured in a temporary exhibition in 2017 about Nobel Prizes called *Nobel Science. Feringa, Zernike and the Groningen tradition*. This exhibition presented not only Nobel Prize laureates, but also shone light on important scientists who never received the prize, but (arguably) should have been awarded one.[54] Nowadays, Zernike and Feringa still have a place in the University Museum in Groningen as part of the permanent exhibition called *Masterminds*.

The Allard Pierson Museum collection of the University of Amsterdam most notably contains the grating used by Zeeman for research leading to the Zeeman effect. Teylers Museum in Haarlem, where Lorentz spent the final part of his career, holds equipment he purchased for his laboratory. After a renovation in 2017, Lorentz's former lab has become the venue for a theatrical performance called "the Lorentz Formula" designed for the museum's visitors.

In 1981 the University Museum Utrecht organized an exhibition about Lorentz called *Hendrik Antoon Lorentz, 1853-1982: Mens en Geleerde*. This exhibition highlighted Lorentz's scientific contributions but less so his status as Nobel Prize laureate. In 2010, the same museum also organized an exhibition about Gerard 't Hooft called *Master the Universe*.[55]

## Acknowledgement

We thank Cis van Heertum for the English corrections.

## Notes

1 The first author has served as curator at Rijksmuseum Boerhaave since 2003. The last sections of this paper partly draw on his own recollections and those of his colleagues. The second author studied Maria Rooseboom's directorship and "Nobel collections" in other Dutch museums as part of her internship in the museum. We thank former curator Floris H. Cohen and Hibrand Wouters for their comments on our first version.

2 On Kamerlingh Onnes and his research: Dirk van Delft, *Freezing Physics: Heike Kamerlingh Onnes and the Quest for Cold* (Amsterdam: Koninklijke Nederlandse Akademie voor Wetenschappen, 2007).

3 Willem Otterspeer, "Begin en context van het Museum Boerhaave," in *75 jaar Museum Boerhaave* (Leiden: Museum Boerhaave, 2006), 5–13; Ad Maas, "Crommelins elitaire instrumenten," *Nederlands tijdschrift voor natuurkunde* 73 (2007), 184–187. On the history of Rijksmuseum Boerhaave, see also Tim Huisman,"National Museum Boerhaave, a Chronical: From Cluttered Attic to One of the World's Top Museums," in *Investigators, Inventors and Innovators: Five Centuries of Research and Innovation in the Netherlands*, edited by Tim Huisman and Ad Maas (Zwolle, Leiden: W Books, Rijksmuseum Boerhaave, 2017), 14–19; Ad Maas, "Introduction: History of Science Museums between Academics and Audiences," *Isis* 108 (2022), 360–265; Tim Huisman, "Layers of Meaning, from Scientific Instrument to Exhibition in Museum Boerhaave," in *Centers of Accumulation in and around the Netherlands during the Early Modern Period*, edited by Lissa Roberts (Berlin: LFT, 2011), 231–254; Dirk van Delft, "Museum Boerhaave en het primaat van de collectie," in *"Wonderen zijn verricht door de geestdrift van de stichters": impressies van een eeuw wetenschaps- en universiteitsgeschiedenis in de Lage Landen*, edited by David Baneke, Astrid Fintelman and Huib Zuidervaart, special issue of *Studium* 6 (2013), 215–233.

4 Klaas van Berkel, *Citaten uit het boek der natuur: opstellen over Nederlandse wetenschapsgeschiedenis* (Amsterdam: Bert Bakker, 1998), 221–239; Klaas van Berkel, "De beoefening van de Nederlandse wetenschapsgeschiedenis in de tweede helft van de negentiende eeuw," *Gewina* 18 (1995), 181–191.

5 Bert Theunissen, *Nut en nog eens nut: wetenschapsbeelden van Nederlandse natuuronderzoekers 1800-1900* (Hilversum: Verloren, 2000), 61.

6 Cited in Van Berkel, "Beoefening," 185.

7 For a classification of science museums, based on their origins, see Lara Bergers and Didi van Trijp, "Science Museum: A Panoramic View," *Isis* 108 (2017), 266–370.

8 This still existing organization for the history of science and medicine is now called Gewina.

9 Bastiaan Willink, *De Tweede Gouden Eeuw: Nederland en de Nobelprijzen* (Amsterdam: Bakker, 1998).

10 Which in fact also belonged to the inventory of the *Natuurkundig Laboratorium*: Zeeman was the man behind the Zeeman Effect (which won him a Nobel Prize) while still an assistant to Kamerlingh Onnes.

11 Being a botanist, a Nobel Prize would not seem an obvious honour for De Vries, but he was nevertheless a Nobel Prize nominee in Physiology or Medicine three times.

12 C.A Crommelin, *Stichting het Nederlandsch Historisch Natuurwetenschappelijk Museum te Leiden. Gids door de Christiaan Huygens verzameling en door de afdeelingen voor natuurkunde, sterrekunde, aardmeetkunde en microscopie* (Leiden: N.V. Boek- en steendrukkerij IJdo, 1931). Italics quotation in the original.

13 Bergers and Van Trijp, "Panoramic View."

14 C.A. Crommelin, "Toespraak gehouden bij de opening van het Nederlandsch Historisch Natuurwetenschappelijk Museum te Leiden den 5den juni 1931," *Physica, Nederlandsch Tijdschrift voor Natuurkunde* 11 (1931), 152–157.

15 Notes made by Crommelin for a speech given on the occasion of the opening of the National Museum for the History of the Sciences, 1947, Archives Rijksmuseum Boerhaave.

16 Crommelin, *Gids door de Christiaan Huygens Verzameling*. In slightly different wordings, Crommelin also mentioned the 'act of piety' in the opening speech for the museum: Crommelin, "Toespraak."

17 Ad Maas, *Atomisme en individualisme: de Amsterdamse natuurkunde tussen 1877 en 1940* (Hilversum: Verloren, 2002), quotation on 206. Underlining in the original.

18 Wilhelmina van Itallie-Van Embden, "Sprekende portretten. Prof. dr. Pieter Zeeman," *Haagsche Post* 5 March 1927.

19 Jeroen Schmale, "Als held kun je maar beter niet uit Nederland komen," *Algemeen Handelsblad* 7 November 2015.

20 Rob van den Berg, *Een gedreven buitenstaander: J.H. van 't Hoff, de eerste Nobelprijswinnaar voor scheikunde* (Amsterdam: Prometheus, 2021), 14–15.

21 Van Berkel, "Beoefening," 191.

22 Crommelin, "Toespraak," 157.

23 This did not seem to have involved a marked change of policy, but it is something that needs to be investigated more thoroughly. General studies we have used for this section include: Leen Dorsman, Ed Jonker and Kees Ribbens, *Het zoet en het zuur, geschiedenis in Nederland* (Amsterdam: Wereldbibliotheek, 2000); *1950. Welvaart in zwart-wit*, edited by C.J.M. Schuyt and Ed Taverne (Den Haag: Sdu Uitgevers, 2000); Jaap Willems, "Van verlichten naar verleiden, inleiding," *Gewina* 23 (2000), 162–164; Lou Dalderup, "Wetenschapsvoorlichting en wetenschapsbeleid in Nederland: 1950-2000," *Gewina* 23 (2000), 165–192. In addition, the annual reports published by the museum have been used extensively.

24 Marian Fournier, "Maria Rooseboom (1909-1978)," *Studium* 6 (2013), 300–303, on 300.

25 Klaas van Berkel, *Dijksterhuis: een biografie* (Amsterdam: Bakker, 1996), 208–220.

26 Ibidem, 414–430 and 450–475.

27 F.J. Duparc and W.A. van Es, *Een eeuw strijd voor Nederlands cultureel erfgoed: Ter herdenking van een eeuw rijksbeleid ten aanzien van musea, oudheidkundig bodemonderzoek en archieven 1875-1975* ('s-Gravenhage: Staatsuitgeverij, 1975), 302; Actually, already in 1935-1936 the museum had been involved in a lecture series at Leiden University meant to bring the humanities and the sciences together, see: Van Berkel, *Dijksterhuis*, 221–222; Otterspeer, "Begin en context," 10–11.

28 Rijksmuseum voor de geschiedenis der natuurwetenschappen, *Verslag van de directrice over het jaar 1950* ('s-Gravenhage: Staatsdrukkerij- en uitgeverijbedrijf, 1950), 1.

29 "Wij spraken met... Mej. dr. Maria Rooseboom," *Het Vaderland* 24 August 1960.

30 Dorsman, Jonker and Ribbens, *Zoet en zuur*, 130–131.

31 Maria Rooseboom, "Musea," in *Vijftig jaren beoefening van de geschiedenis der geneeskunde, wiskunde, natuurwetenschappen in Nederland*, edited by B.P.M. Schulte (n.p. 1963), 89–107.

32 This correspondence also included the famous telegram Lorentz sent to Einstein on September 22th 1919 about the success of Arthur Eddington's solar eclipse expedition to measure the bending of starlight by the sun, a sensational confirmation of Einstein's Theory of General Relativity.

33 *Albert Einstein & Museum Boerhaave*, edited by Peter de Clercq and Anne C. van Helden [Mededeling 250] (Leiden: Museum Boerhaave, 1993), 9–11.

34 "Tentoonstelling Lorentz-Kamerlingh Onnes, twee grote geleerden en oude vrienden op treffende wijze geëerd," *Algemeen Handelsblad* 20 June 1953.

35 Maria Rooseboom, *Tentoonstelling H. A. Lorentz en H. Kamerlingh Onnes, 1853-1953: catalogus* (Leiden: Rijksmuseum voor de Geschiedenis der Natuurwetenschappen, 1953).

36 "Lorentz en Kamerlingh Onnes een eeuw geleden geboren," *Nieuwe Haarlemsche Courant*, 25 June 1953.

37 "Grootste snelheden en laagste temperaturen brachten de Nederlandse wetenschap wereldfaam," *Het Vrije Volk*, June 27, 1953.

38 *Rijksmuseum voor de geschiedenis der natuurwetenschappen. Verslag van de directrice over het jaar 1952* ('s-Gravenhage: Staatsdrukkerij- en uitgeverijbedrijf, 1952); *Chemisch Weekblad, orgaan van de Nederlandse chemische vereniging* 30 August 1952.

39 *Verslag van de directrice over het jaar 1953*.

40 *Verslag van de directrice over het jaar 1956, 1957 and 1966*.

41 For an impression: *Museum Boerhaave. De inrichting van het Caecilia-Gasthuis te Leiden* (n.p. 1991).

42 For this period: interview held by Ad Maas with former museum staff members Gerrit Veeneman, Marian Fournier and H.Floris Cohen on March 21, 2018; interview held by Tim Huisman with Bernie Gogelein (widow of former director Dolf Gogelein), on December 14, 2018.

43 Dorsman, Jonker and Ribbens, *Zoet en zuur*, 85–127, quotations on 87 and 124.

44 Cf. Van Berkel, "Beoefening," 191.

45 Bastiaan Willink, "Een inleiding tot de Tweede Gouden Eeuw: de wetten van 1863 en 1876 en de wedergeboorte van de Nederlandse natuurwetenschap," *Hollands Maandblad* 22 (1980), 3–9.

46 https://nos.nl/artikel/2451576-stem-van-nederlandse-nobelprijswinnaar-klinkt-na-bijna-100-jaar (accessed November 17, 2022).

47 Dirk van Delft, Mieneke te Hennepe and Esther van Gelder, *Nobelprijs NL. Eregalerij van Nederlandse Nobelprijswinnaars* (Leiden: Museum Boerhaave, 2011).

48 Rogier van Reekum, *Out of Character. Debating Dutchness, Narrating Citizenship* (PhD thesis Erasmus University Rotterdam, 2014), esp. 134–144, and 209–215.

49 Ibidem.

50 *Jaarverslag Rijksmuseum Boerhaave 2017*, https://rijksmuseumboerhaave.nl/over-ons/organisatie/jaarverslagen/, 9 and 19.

51 "The Nobel Prizes and the Public Image of Science," special issue of *Public Understanding of Science* edited by Sven Widmalm 27 (2018).

52 Lisa Kirch, "The Changing Face of Science and Technology in the Ehrensaal of the Deutches Museum, 1903-1955," [Deutsches Museum preprint, issue 13, 2017], quotation on 21.

53 Exhibition guide: *Frits Zernike, Groninger Nobelprijsdrager 1888-1966*, edited by H. Brinkman (Groningen, Amsterdam: Universiteitsmuseum Groningen, North-Holland Publishing Company, 1988).

54 For more information, news reports, videos and the exhibition e-guide, visit the museum site: https://www.rug.nl/museum/exhibitions/previous/2017/ (accessed November 17, 2022).

55 UMU Projectplan, het onderzoeksmuseum van Nederland (Universiteitsmuseum Utrecht 2021).

## Bibliography

### *Primary sources*

### Rijksmuseum Boerhaave Archive

Rooseboom, Maria. *Tentoonstelling H. A. Lorentz en H. Kamerlingh Onnes, 1853-1953: catalogus.* Leiden: Rijksmuseum voor de Geschiedenis der Natuurwetenschappen, 1953.

Notes made by Crommelin for a speech given on the occasion of the opening of the National Museum for the History of the Sciences, 1947.

Interview by Tim Huisman with Bernie Gogelein (widow of former director Dolf Gogelein), November 14, 2018.

Interview by Ad Maas with former museum staff members Gerrit Veeneman, Marian Fournier and H. Floris Cohen, March 21, 2018.

### Newsarticles

*Algemeen Handelsblad,* "Tentoonstelling Lorentz-Kamerlingh Onnes, twee grote geleerden en oude vrienden op treffende wijze geëerd." June 20, 1953.

*Chemisch Weekblad, orgaan van de Nederlandse chemische vereniging.* August 30, 1952.

*Het Vaderland,* "Wij spraken met... Mej. dr. Maria Rooseboom." August 24, 1960.

*Het Vrije Volk,* "Grootste snelheden en laagste temperaturen brachten de Nederlandse wetenschap wereldfaam." June 27, 1953.

*Nieuwe Haarlemsche Courant,* "Lorentz en Kamerlingh Onnes een eeuw geleden geboren." June 25, 1953.

*NOS*, "Stem van Nederlandse Nobelprijswinnaar klinkt na bijna 100 jaar." November 8, 2022. https://nos.nl/artikel/2451576-stem-van-nederlandse-nobelprijswinnaar-klinkt-na-bijna-100-jaar.

## Other primary sources

Universiteitsmuseum Utrecht, *UMU Projectplan, het onderzoeksmuseum van Nederland*, 2021. https://umu.nl/de-vernieuwing/.

University Museum Groningen, *Exhibition guide: Frits Zernike, Groninger Nobelprijsdrager 1888-1966*, edited by H. Brinkman. Groningen: University Museum Groningen, 2017. https://www.rug.nl/museum/exhibitions/previous/2017/?lang=en.

## *Secondary sources*

Berg, Rob van den. *Een gedreven buitenstaander: J.H. van 't Hoff, de eerste Nobelprijswinnaar voor scheikunde*. Amsterdam: Prometheus, 2021.

Bergers, Lara and Didi van Trijp. "Science Museums: A Panoramic View," *Isis* 108 (2017), 266–370.

Berkel, Klaas van. *Citaten uit het boek der natuur: opstellen over Nederlandse wetenschapsgeschiedenis*. Amsterdam: Bert Bakker, 1998.

Berkel, Klaas van. *Dijksterhuis: een biografie*. Amsterdam: Bert Bakker, 1996.

Berkel, Klaas van. "De beoefening van de Nederlandse wetenschapsgeschiedenis in de tweede helft van de negentiende eeuw," *Gewina* 18 (1995), 181-191.

Clercq, Peter de and Anne C. van Helden, eds. *Albert Einstein & Museum Boerhaave* [Mededeling 250]. Leiden: Museum Boerhaave, 1993.

Crommelin, C.A. "Gids door de Christiaan Huygens verzameling en door de afdeelingen voor natuurkunde, sterrekunde, aardmeetkunde en microscopie," *Stichting het Nederlandsch Historisch Natuurwetenschappelijk Museum te Leiden*. Leiden: N.V. Boek- en steendrukkerij Udo, 1931.

Crommelin, C.A. "Toespraak gehouden bij de opening van het Nederlandsch Historisch Natuurwetenschappelijk Museum te Leiden den 5den juni 1931," *Physica, Nederlandsch Tijdschrift voor Natuurkunde* 11 (1931), 152–157.

Dalderup, Lou. "Wetenschapsvoorlichting en wetenschapsbeleid in Nederland: 1950-2000," *Gewina* 23 (2000), 165–192.

Delft, Dirk van. *Freezing Physics: Heike Kamerlingh Onnes and the Quest for Cold*. Amsterdam: Koninklijke Nederlandse Akademie, 2007.

Delft, Dirk van. "Museum Boerhaave en het primaat van de collectie," in *"Wonderen zijn verricht door de geestdrift van de stichters": impressies van een eeuw wetenschaps- en universiteitsgeschiedenis in de Lage Landen*, edited by David Baneke, Astrid Fintelman and Huib Zuidervaart, 215–233. The Hague: Gewina, 2013.

Delft, Dirk van, Mieneke te Hennepe and Esther van Gelder. *Nobelprijs NL. Eregalerij van Nederlandse Nobelprijswinnaars*. Leiden: Museum Boerhaave, 2011.

Dorsman, Leen, Ed Jonker and Kees Ribbens. *Het zoet en het zuur, geschiedenis in Nederland*. Amsterdam: Wereldbibliotheek, 2000.

Duparc, F.J. and W.A. van Es. *Een eeuw strijd voor Nederlands cultureel erfgoed: Ter herdenking van een eeuw rijksbeleid ten aanzien van musea, oudheidkundig bodemonderzoek en archieven 1875-1975*. The Hague: Staatsuitgeverij, 1975.

Fournier, Marian. "Maria Rooseboom (1909-1978)," *Studium* 6 (2013) 300–303.

Huisman, Tim. "Layers of Meaning, from Scientific Instrument to Exhibition in Museum Boerhaave," in *Centers of Accumulation in and around the Netherlands during the Early Modern Period*, edited by Lissa Roberts, 231-254. Berlin: LFT, 2011.

Huisman, Tim. "National Museum Boerhaave, a Chronical: From Cluttered Attic to One of the World's Top Museums," *Investigators, Inventors and Innovators: Five Centuries of Research and Innovation in the Netherlands*, edited by Tim Huisman and Ad Maas, 14–19. Zwolle, Leiden: WBooks, Rijksmuseum Boerhaave, 2017.

Itallie-van Embden, Wilhemina van. "Sprekende portretten. Prof. dr. Pieter Zeeman, *Haagsche Post* 5 (1927).

Kirch, Lisa. "The Changing Face of Science and Technology in the Ehrensaal of the Deutches Museum, 1903-1955," [Deutsches Museum preprint issue 13] (2017).

Maas, Ad. *Atomisme en individualisme: de Amsterdamse natuurkunde tussen 1877 en 1940*. Hilversum: Verloren, 2002.

Maas, Ad. "Crommelins elitaire instrumenten," *Nederlands tijdschrift voor natuurkunde* 73 (2007), 184–187.

Maas, Ad. "Introduction: History of Science Museums between Academics and Audiences," *Isis* 108 (2022), 360–265.

Otterspeer, Willem. "Begin en context van het Museum Boerhaave," in *75 jaar Museum Boerhaave*, 5–13. Leiden: Rijksmuseum Boerhaave, 2006.

Reekum, Rogier van. *Out of Character. Debating Dutchness, Narrating Citizenship*. Rotterdam: PhD thesis Erasmus University, 2014.

Rijksmuseum voor de geschiedenis der natuurwetenschappen. *Verslag van de directrice over het jaar 1950*. The Hague: Staatsdrukkerij- en uitgeverijbedrijf, 1950.

Rijksmuseum voor de geschiedenis der natuurwetenschappen. *Verslag van de directrice over het jaar 1952*. The Hague: Staatsdrukkerij- en uitgeverijbedrijf, 1952.

Rijksmuseum voor de geschiedenis der natuurwetenschappen. *Verslag van de directrice over het jaar 1953*. The Hague: Staatsdrukkerij- en uitgeverijbedrijf, 1953.

Rijksmuseum voor de geschiedenis der natuurwetenschappen. *Verslag van de directrice over het jaar 1956*. The Hague: Staatsdrukkerij- en uitgeverijbedrijf, 1956.

Rijksmuseum voor de geschiedenis der natuurwetenschappen. *Verslag van de directrice over het jaar 1957*. The Hague: Staatsdrukkerij- en uitgeverijbedrijf, 1957.

Rijksmuseum voor de geschiedenis der natuurwetenschappen. *Verslag van de directrice over het jaar 1966*. The Hague: Staatsdrukkerij- en uitgeverijbedrijf, 1966.

Rijksmuseum Boerhaave. *Jaarverslag Rijksmuseum Boerhaave 2017*, 2017. https://rijksmuseumboerhaave.nl/over-ons/organisatie/jaarverslagen/.

Rooseboom, Maria. "Musea," in *Vijftig jaren beoefening van de geschiedenis der geneeskunde, wiskunde, natuurwetenschappen in Nederland*, edited by B.P.M. Schulte, 89–107. N.p., 1963.

Schmale, Jeroen. "Als held kun je maar beter niet uit Nederland komen," *Algemeen Handelsblad*, November 7, 2015.

Schuyt, C.J.M. and Ed Taverne, eds. *Welvaart in zwart-wit*. The Hague: Sdu Uitgevers, 2000.

Theunissen, Bert. *Nut en nog eens nut: wetenschapsbeelden van Nederlandse natuuronderzoekers 1800-1900*, Hilversum: Verloren, 2000.

Widmalm, Sven, ed. "The Nobel Prizes and the Public Image of Science," special issue of *Public Understanding of Science* 27 (2018), 390–494.

Willems, Jaap. "Van verlichten naar verleiden, inleiding," *Gewina* 23 (2000), 162–164.

Willink, Bastiaan. *De Tweede Gouden Eeuw: Nederland en de Nobelprijzen*. Amsterdam: Bakker, 1998.

Willink, Bastiaan. "Een inleiding tot de Tweede Gouden Eeuw: de wetten van 1863 en 1876 en de wedergeboorte van de Nederlandse natuurwetenschap," *Hollands Maandblad* 22 (1980), 3–9.

CHAPTER FOUR

# What Heroes does Literature Need? Insight into the Nobel Prize as a Literary Motif

*Daniela Link*

**Abstract**

As quick as one is willing to call someone a "hero," it is just as complicated to grasp this term in literature. That's why a distinction is often made between subcategories such as classic heroes, antiheroes or superheroes – they are self-sacrificing, clever or magical figures, but one thing unites them: they are not Nobel laureates. As a literary motif, the Nobel Prize radiates as an expression of genius, excellence, honor and yes, also a heroic sacrifice of the spirit of research. It thus represents a Faustian striving, legitimized by the success of the honor. Using selected examples, the article examines the function of the (Nobel) Prize as a literary motif and asks to what extent the function of the motif is reduced to the depiction of excellence and science. Which breaks in the depiction and which image of science and society are conveyed via the motif and what influence does this have on the characters?

**Keywords:** Hero, Nobel Prize, genius, literary motif

This article examines the Nobel Prize and the status of its laureates in the mirror of literature. The Nobel Prize motif is found in a broad spectrum of literature: from the classic light novel to shallow thrillers, and from sophisticated novellas to comics. The study starts with the assumption that the Nobel Prize not only functions as a sign of excellence, but that it also reflects social and political discourses.

Many literary texts that use the Nobel Prize motif thematize the scientific sector and exclude the field of literary prizes. A similar observation has already been made in a study of the Nobel Prize as a motif in film and television.[1] As a result, the Nobel Prize in Literature is largely ignored as a subject of literary representation. Even when it is mentioned, as in Paul Ernst's 1919 novella *The Nobel Prize*, it is used merely as a tongue-in-cheek autobiographical bon mot. Ernst's novella refers self-deprecatingly to the author's disappointed hopes for the Nobel Prize and concludes that the laureate of the Nobel Prize for Literature received it because of relationships and friendships. Therefore, this study also refers to texts that deal with the Nobel Prizes in science.

This paper examines how the Nobel Prize is used as a motif and what function it assumes in German and English fictional texts. The focus will be on concrete textual analysis, rather than on literary-historical classification. Since genius is a quality that is often attributed to Nobel Prize laureates, this article first presents this elusive and contingent concept in its various manifestations.

## Heroes of science?

### *Hero of knowledge*

The idea of the genius is related to the much older heroes-motive. Characteristically, the hero is the protagonist of a literary text; he or she is the plot carrier who takes on representative functions. Literary heroes, figures like Odysseus, Arthur, or even Aragorn, are active fighters, symbols of morally good action and saviours in distress. The term is therefore not value-neutral per se but arouses a particular expectation in the reader.[2]

Additionally, even the early heroes of the Greek mythological world reveal that other characteristics are involved as well. All literary heroes act independently of the masses and take special paths that lead them to their goal. Non-conformism and dissociation can thus be identified as signs of heroism.[3] Younger heroes in particular display these characteristics, like Thomas Mann's protagonist Adrian Leverkühn in *Doktor Faustus*, whose pursuit of perfection and musical innovation turns him into a lonely, hunted and half-mad man. Particularly these characteristics: the loneliness, the avoidance of the masses, the willingness to self-sacrifice for the sake of a dreamed goal establish the connection between literary heroic figures and the idea of the genius, who, as a thinker and researcher, does not really fit the image of the man of action.

The genius possesses exceptional creative, formative or scientific abilities and unlike others is able to realize his goals and ideas consistently.[4] Like the literary hero, the genius appears as the committed rescuer and fighter for the good cause.

It was at the beginning of the twentieth century that the notions of genius and heroes coalesced. This era heralded the emergence of mass movements. Politicians, scientists, psychologists, and philosophers realized with horror that these were becoming a factor of influence. A pioneer of mass theories, the physician and psychologist Gustave le Bon (1841-1931), described in his best-known work, *Psychologie des foules* (1895), masses as forces of destruction. Characteristic for mass members was the loss of their individuality. They were dependent on leadership, strongly emotionalized and acted irrationally.[5] Le Bon explicitly uses the medical term "contagion"[6] to describe this unstoppable process.

The genius in this period is staged as a counter-image of this mass-member, as a hero who is seen by the old elites as an embodiment of resistance against the forces of modernity. In this context, geniuses replaced the classical hero figures and established themselves as saviours of the old order.[7] The new heroes represented the rulers, the white, well-off men: researchers, statesmen, artists. They became symbols and icons of their nation and were accordingly politicized and included in international competition for other purposes.[8]

*History of the genius*

Earlier, in the Sturm und Drang period, geniality was still considered a characteristic of the artist and a subject for artists, as Goethe's *Prometheus*, for example, shows. It was largely characterized by a special closeness to nature, an intense emphasis on emotion and a particular intuition. From the 1850s onwards, it became a topic of scientific interest, particularlyof the newly emerging field of psychiatry, heredity, and the natural sciences in general, inducing a paradigm shift around 1900. Genius left the space of the creative-artistic. Besides artists, also political leaders, philosophers, discoverers, and scientists who were awarded the title of "genius."[9] A veritable genius frenzy followed, and during this frenzy, the first Nobel Prize was awarded in 1901. From then on, the idea of genius was combined with the category of intelligence and the nimbus of special privileges. Genius reached science and researchers became heroes.

*Genius and madness*

Geniuses are the heroes of the early twentieth century and the Nobel Prize is the crowning glory of their career, the insignia of their genius. However, as early as the Romantic period, another genius has also conquered the world stage and shaped its own stereotype: the mad genius. In German-language literature, it was above all the genius-mad characters of E.T.A. Hoffmann (1776-1822) who shaped the image of the mad artist or researcher. Radical self-dereliction leads them to the loss of a sense of reality and ultimately to madness; they represent outsiders and eccentrics.[10] Only in seclusion, separated from mass trends, can they live out their special abilities. But this special status is not only a privilege. Especially in recent literature, it has been increasingly transformed into the image of a lost individual suffering in loneliness, whose proximity to madness is always present. The ambivalence of the portrayal creates the amoral-looking anti-hero who, in the worst case, is driven by a lust for power and domination, crossing the line between genius and madness.[1112] Genius – at least from a literary point of view – carries the risk of madness.

In addition to the portrayal of the sick, power-hungry scientist, the mad genius has in recent years undergone a reinterpretation in the guise of the autistic scientist. Ingenious, but separated from the social world due to his or her neurodivergent perception, this stereotype was solidified with *Rain Man* (1988) and provided recipients of literature and film with a whole series of highly gifted researcher figures, whose affiliation to the autistic spectrum is open to wild speculation. Retrospective classifications of historical geniuses such as the 1922 Nobel Prize laureate Albert Einstein also link the literary figure of the autistic genius to the representation of genius scientists and Nobel Prize laureates.[13]

Genius is therefore a mutable construct that changes over time, social attitudes and political mood. Geniuses can be the self-sacrificing saviours of humanity, but they can also be power-hungry, maniacal despots. Consequently, the image of the genius is just as diverse as that of the Nobel Prize laureate, as the following analysis of several selected texts will reveal.

## The Nobel Prize as a literary motif

### *Nobel laureates from the perspective of the development novel*

A recent example from the German-speaking world that explicitly refers to Nobel Prize laureates, is the novel *Fast genial* (2011) by Benedict Wells. The story combines the characteristics of the coming-of-age novel with those of a road novel. Wells' novel describes how the search for his father becomes a trip of self-discovery for the young Francis with a surprising outcome. Following the debates of early twentieth-century researchers and the question of the genes of geniuses, the novel discusses the possibilities and, above all, the limits of breeding geniuses. Thus, the protagonist himself is seemingly the product of a gene bank of genius. The preservation and transmission of genius is the main purpose of this sperm bank. Its founder strives to strengthen intelligence and fears the increase of "stupid" children. Nobel Prize laureates are explicitly recruited for this purpose:

> Mr. Monroe sagte, dass der intelligente Mensch auszusterben drohe. Während die Akademiker sich nicht mehr richtig vermehren, sondern nur noch Karriere machen würden und gerade Genies oft ohne Nachkommen seien, würden die Dummen ein Kind nach dem anderen in die Welt setzen. Dem wolle er etwas entgegenstellen. Er habe deshalb Unsummen ausgegeben, um die Samen von diversen Nobelpreisträgern und genialen Wissenschaftlern zu kaufen.[14]

The fortune sperm donations of Nobel Prize laureates cost also brings the material value of genius into focus.[15]

The Nobel Prize becomes a guarantor of excellent hereditary qualities and a symbol of special qualities worthy of preservation, and fuels the old belief in a new and stronger human race, as would correspond to Nietzsche's *Übermensch*. This idea, dangerous from a historical point of view, is negated within the novel by means of ironic twists. The genius breeding centre subsides not only because the leading researcher is confronted with his alleged past in National Socialism, but also because the results of his experiments do not meet the expectations. The genius rate is low: "Alle anderen Kinder waren eher durchschnittlich. Und dann gab es sogar noch Nieten. Mit völlig unverständlich. Man nimmt die besten Zutaten und züchtet Versager."[16] The research is therefore unethical and the result questionable. At the end of the novel, all that remains is the observation that the genes of Nobel Prize laureates are no guarantee of success.

Following the Romantic tradition, Wells' novel places itself in line with the horror stories of Hoffmann or Shelley's *Frankenstein* (1818). At the same time, it ironically negates the idea of a mad genius alchemist breeding a new human race.

The novel breaks with the expectations of the recipients. Three sperm donors are presented explicitly or implicitly. The first is the protagonist's father, once the institute's most coveted donor, as the novel progresses, it turns out that he is not a genius from Harvard, but a homeless fraud.[17] The success of another donor is also questionable, as Alistair, who was once praised as a child genius, is just as gifted as his genius father, but is now a young man who is failing to cope with the realities of life and keeps his head above water with the help of cannabis. The only indirectly mentioned Nobel Prize laureate does not embody the character traits of a hero either; he ruthlessly rejects and denies his child conceived by a sperm bank. Humanity, sincerity and empathy are not the characters' strong points. The Nobel Prize is used by Wells as a symbol of desirable excellence, only to be negated on all levels of the novel. Ultimately, the eternal striving and longing for success reveal that genius cannot be forced. Inherited or bred genes remain a material for utopias. The hero of the novel in the end learns to find his way to more self-confidence and a new future even without genius genes. Wells uses the radiance of the Nobel Prize above all as a foil for contrast.

### *The Nobel Prize in entertainment literature – between excellence and gender struggle*

The aura of the Nobel Prize extends to the selection process and the award itself, which resembles a coronation ceremony. In 2005, the Nobel Foundation, the selection of the laureate and the final award ceremony became the setting for a thriller. Andreas Eschbach's entertainment novel *Der Nobelpreis* seemingly traces the scenery of a corrupt Nobel system that is purely geared to economic interests and power. At first glance, the reader is presented with the following plot: a member of the Nobel Committee, a professor, is blackmailed into voting in favour of a certain researcher in the election for the Nobel Prize in physiology or medicine. The kidnapping of his daughter and the subsequent search for the girl takes place in parallel with the selection process. The plot ends shortly after the official awarding of the prize. The author uses the weeks of narrative time to detail the entire process, the ceremonial procedures and finally the celebratory banquet. Eschbach's novel offers recipients a behind-the-scenes look at the awarding of the Nobel Prize. *The Nobel Prize* suggests, as the following quote underlines, the corrupt nature of the Nobel Prize committee; collusion, venality, fraud, and competitiveness characterize the scene:

> Ab und zu ein Gefälligkeitsgutachten, okay, aber das Sparkonto in der Schweiz wächst, also was soll's? Eine kleine Auskunft unter Freunden? Scheiß auf die Vorschriften, wenn der nächste Vortrag in Thailand stattfindet." Bosse hielt inne, nickte sinnend. "Alles Kinderkram natürlich. Im Grunde war mir von Anfang an klar, dass die langfristig vorhatten, einen Nobelpreis zu kaufen."[18]

Eschbach describes a system of mutual favours, pronounced campaigning and targeted intervention by pharmaceutical companies in the awarding of the Nobel Prize. The image of the Nobel Prize is consistently dismantled. It is no longer individual genius that paves the way to the Nobel Prize; instead, contacts and unscrupulous benefactors are the decisive forces. The novel exaggerates the real-life campaigning[19] that tries to recommend certain people for the Nobel Prize. Murder, kidnapping, blackmail and threats are the means of choice. But, as is usual in entertainment literature, the novel breaks up this picture and exposes the apparently blackmailed professor, the member of the Nobel Committee, as the initiator of an absurd campaign, a private feud in which the Nobel Prize serves merely as a prop. In this context, the Nobel Prize is an example of power and influence, but it also remains a prop that is misappropriated by the actions of the mad professor.

This entertainment thriller furthermore addresses another topic: the issue of gender. No other text discussed here highlights the discrepancy between female and male scientists and the unbalanced relationship within the research-community and the Nobel Committee so clearly. The number of women on the Nobel Committee is small, women in science are devalued and not taken seriously:

> Es ist mir auch nicht entgangen, dass die meisten abfällig über Sofía Hernández Cruz sprechen. Weil sie eine Frau ist. Eine Spanierin zumal – unvorstellbar, dass so jemand bedeutende Arbeit auf dem Gebiet der Neurophysiologie leisten kann, nicht wahr? Ganz zu schweigen von dieser unsäglichen Moraldiskussion. So sind sie, die Vorurteile unserer geschätzten Kollegen.[20]

The novel, however, belies the disparagement of women scientists addressed in the quote above. The value of the Nobel Prize and the integrity of the committee are restored and the well-deserved honour of the Nobel laureate, Hernández Cruz, takes place. Not only that, she solves the mystery of the novel and exposes the intrigue within minutes:

> Diese Frau verdient den Nobelpreis, wenn ihn irgendjemand verdient. Sie hat ein paar Minuten nachgedacht und dann genau die eine Frage gestellt, die ich mir schon seit zwei Wochen hätte stellen sollen, auf die ich aber einfach nicht gekommen bin. Die eine Frage, die alles zerlegt, was ich geglaubt habe, und es neu zusammensetzt.[21]

The Nobel Prize laureate solves the mystery and once again proves her excellence and genius. With this successful example of a successful woman this entertaining novel seems to break with common gender stereotypes. Nevertheless, if you consider not only the plot level but also the structure and language, the picture is reversed. For Eschbach's Nobel novel presents women, especially the Nobel laureate, as sexual objects. It is primarily the visual appearance and the erotic voice that are used to characterize the laureate. Noteworthy, in this context is the absence of the Nobel laureate. In contrast to the members of the Nobel Committee, she never makes an appearance; her statements are made in the form of television interviews or through the reproduction of male third parties. Eschbach's novel takes away the voice of the very woman who receives the insignia of genius and excellence, who is also entitled to claim them, and makes her appearance exclusively through male messengers. All these references confirm, on a meta-level, the ongoing gender conflict and the paternalism that women face in academia. The novel paints a very stereotypical cliché: successful women must have erotic appeal, charm and seductive skills. This cliché regarding female genius and success can be related to the tradition of the femme fatale. The concepts of femme fatale and femme fragile have codified men's desires and fears since the fin de siècle. As a demonic seductress, the femme fatale embodies the terrifying and alluring realm of female sexuality. While the delicate, Madonna-like and exalted femme fragile represents an unattainable male ideal.[22] This schema can also be related to the targeted performance of Hernández Cruz, who allegedly uses her femininity to advance her career. The danger of degrading women is ever-present in this image, and so the stereotype ultimately serves (in keeping with the genre) mainly to preserve familiar clichés.

A more recent novel traces the social and political relevance of the Nobel Prize from a more contemporary perspective: the currently very topical narrative of a pandemic that is spreading around the world. Christine Sweeney-Baird's *The End of Men* (2021) is another dystopia. It depicts the development and fight against a disease that wipes out the majority of the world's male population in the form of several loosely connected and repeatedly recurring individual fates. The parallels to the worldwide fight against Covid are obvious and so the fictional developers of the vaccine also receive the Nobel Prize: "I'm going to win the Nobel Prize. Of course I am. Everybody says so. It's the first time the Swedes are awarding them since the whole world went to shit, only in three categories – Physiology or Medicine, Chemistry and Peace – and I'm a shoo-in."[23] At this point, the Nobel Prize appears to be the rightful reward for outstanding scientific achievement. Yet the Nobel laureate, Dr Lisa Michael, is not a classic heroic figure, because it is not pure self-sacrifice or charity that makes her research tireless; career ambition and profit-seeking drive the scientist as well:

> Wendy rushes back in and thrusts her phone at me. Now is not the time for the Canadian Public Health Agency to put me on hold; they'll have been praying I would call.
> 'Lisa,' the voice on the other end of the line says.
> 'Dr Michael's fine,' I reply. I've never spoken to this woman before, we're not on first name terms.
> 'Apologies, Dr Michael. What can I do for you?'
> My voice is bubbling with happiness. 'You should sound more excited to hear from me. This is the phone call that's going to change your life.'
> There's a stunned pause. I can just imagine this woman thinking no, no, surely not. 'Yes, I am in fact a god. I have a vaccine. A hundred per cent success rate. Blood tests have come back clean. We've bypassed the missing chromosomes. I have cured the Plague.'
> 'Dr Michael, I, don't—'
> 'You don't know what to say? Yes, I thought that might be the case. Before you get too excited, there's going to be a difficult conversation between you, me and the Canadian Government.'
> The woman sounds flustered. I can imagine her wearing a blazer at her nice desk in her nice office in her nice comfy job.
> 'What are you talking about?' she asks.
> 'I'm going to sell Canada the vaccine.'
> 'Very funny, Dr Michael.'
> 'I'm not kidding. If you want it, you'll have to pay for it.'
> 'Lisa, Dr Michael. You can't sell the Government a vaccine. It's ... you ... you're a doctor.'[24]

Dr. Lisa Michael is not a sympathetic figure, but the type of the solitary, power-interested and success-oriented researcher. She sells the vaccine despite the worldwide rampant pandemic and thus refutes the image of the self-sacrificing heroine. She is a genius who puts the heroism of the Nobel Prize laureates into perspective:

> 'My name is Ingrid Persson. I am the Chair of the Nobel Assembly at the Karolinska Institute.'

> Oh my God. Oh my fucking God. This is the coolest phone call of my entire life.
> 'I'm thrilled to inform you that we have chosen to award you the Nobel Prize for Physiology or Medicine.'
> 'Thank you! This is an honour, truly. (...)' 'I have another piece of news you might find less ... pleasing.'
> My heart drops. What is it? Maybe it's the money, I don't care about the money. I don't need prize money. No ceremony maybe? Damn it, I've been dreaming about the ceremony my whole life.
> 'You will share the prize.' Ingrid says more words at this point but the world goes a little fuzzy and black around the edges and Margot is looking up at me quizzically and did she just say I have to share the Nobel Prize? I've never even shared an *office*.[25] (emphasis in the original)

In the midst of a global crisis, above all the competitive spirit distinguishes the future Nobel laureate. Lisa Michael's thought speech depicts her as a hard-hearted, egotistical, and eccentric figure. The medical integrity expected of outsiders finds no equivalent here. Once again, a Nobel Prize laureate is endowed with negative character traits, which are emphasized by other characters: "Well I don't envy her. I don't envy her callousness. I envy her bank balance."[26] Women in science lose some of their humanity in this novel as soon as they are portrayed in direct contact with the Nobel Prize. The stereotype of the tough and empathy-less female scientist is deployed, who, intent on her advantage, acts only to her own advantage. The sympathetic figures are the female marginal figures, the losers of the pandemic, the emotionalized and self-sacrificing women. In the two novels discussed in this section, various associated stereotypes like the pursuit of power, lack of empathy, and gender roles are presented and discussed based on the Nobel Prize.

## Age and genius

The previous sections have dealt intensively with the portrayal of the Nobel Prize and the laureates at the zenith of their power and success. But what happens to Nobel Prize laureates when their genius wanes. Irene Dische provides a particularly drastic example in her novella *Der Doktor sucht ein Heim.*[27] The text is a retrospective look at the life of a nameless doctor. The tangled, incomplete narrative, marked by leaps in time, is written in the first-person perspective. The recipients experience the drifting of the ageing Nobel Prize laureate, his Alzheimer's-related memory lapses, aggressive attacks and loneliness in a particularly intense way. The Jewish doctor, born in Drohobytsch, today's Ukraine, and raised in Vienna, had to flee to America when the Nazis marched in, where he worked as a successful chemist and later received the Nobel Prize. This is how far the recipient can reconstruct the course of his life from the fragments of the first-person narrator's memories. Dische's first-person narrator is erratic and hardly reliable due to his progressive illness, of which the reader does not learn this until later in the book. At the beginning, erroneous indications of place mark the problematic nature of the narrative instance: "Der Hudson River trennte mein Haus von ihrem Haus in Drohobyc[28]."[29] The narrator mixes and confuses places and people. Due to his illness, he can hardly be expected to give clear information. He still remembers parts of the Nobel Prize, but the insignia of his genius has slipped away:

> Die anderen Professoren aber respektieren mich, weil ich viele Preise erhalten habe. Vor allem einen, einen kolossal noblen, den jeder haben wollte. Zescha war bei mir, als er mir unter sehr heißen, sehr hellen Scheinwerfern in die Hand fiel. Ich weiß nicht, was ich mit ihm gemacht habe. Ich hatte ihn immer in meinem Schreibtisch. Aber was ist nur aus meinem Schreibtisch geworden? Bestimmt hat ihn jemand gestohlen, hat das Holz zu Haus im Kamin verheizt, und dabei ist die Medaille geschmolzen.[30]

In a childlike tone of voice, the doctor describes his memories of the honour; it is not the value of the award that is relevant, but secondary physical sensations he had at the ceremony: heat and blinding light. Zescha, his sister murdered by the Nazis, accompanies him through his illness. She is a shadow, his guilty conscience and a memory that keeps him company in his loneliness. Thus, in his memory, she is also present at the award ceremony. But it is not only the award ceremony that is described in a childishly naïve manner; the path to the prize also appears easy and playful from the award winner's sketchy memory; like a ball, the Nobel Prize falls to him seemingly effortlessly. This description could be humorous if the comedy did not tip in the face of illness and leave a sobering conclusion:[31] Even genius heroes of science can lose their abilities and not avoid going through the same forgetting process as do all dementia sufferers. Accordingly, the text is not on genius, but on the symptoms and effects of the progressive disease: the permanent worry of being robbed, which is also supposed to explain the loss of his desk, namely, his office and the award. The confusion of places and people, addressing the dead and not recognizing his own daughter are also sad markers of the disease. The once successful scientist becomes a ridiculed old man: "Alle grüßen mich mit größtem Respekt, sie sagen 'Hallo, Doktor' and some say 'Hi, Doktorchen.'"[32] The hero of science has become a factotum in his former place of work. Dische employs means of disparaging comedy, such as rigidity and repetition, to mark the loss of intellectual clarity:

> Dieses Kaffeehaus macht sich für mich wahrhaft bezahlt. Ich habe das Plastikbesteck und die Pappteller auf und benutze sie zu Hause. Ich habe schon Pappservierten genug, um

> allen Mitgliedern der Akademie Cocktail-Happen zu servieren. Auf den Servietten steht MacDonalds – komischerweise. Der Dr. MacDonald, der 1942 den Preis für Chemie gewann, wird denken, es seien seine und ich hätte sie ihm gestohlen.[33]

The connection of the Nobel motif with everyday experiences, the degradation of scientific consecration and the fact that readers experience the feeling of knowing more, of having a clearer perspective than the ageing Nobel laureate, also provides the feeling of superiority.

Instead of a genius, appreciated and revered by colleagues, family and students, Dische's *Der Doktor sucht ein Heim* traces the life of a self-centred, ruthless man who sacrificed his family to pursue a career and left them behind in the face of the Nazi threat. This Nobel Prize laureate is not a hero and never was, but an egoistic and strategically thinking man who, now marked by illness, has also lost his last connection, his research work. Dische's text thus considers the value of the prize in relative terms. In the end, it has brought the doctor no happiness, his former wife increasingly refuses contact and his daughter takes him to an old people's home.

Aldous Huxley's novella *The Genius and the Goddess* is another example of an ageing genius. It looks back on the life of a well-known Nobel Prize laureate and his family, described at the beginning with the help of a photograph:

> It showed three adults standing in front of a wooden summerhouse – a small, thin man with white hair and a beaked nose, a young giant in shirt sleeves and, between them, fair-haired, laughing, broad-shoulders and deep-bosomed, a splendid Valkyrie incongruously dressed, in a hobble skirt.[34]

Huxley's novella describes the connection between eroticism and genius as an interaction. Katy Maartens, the woman, is described as a beauty whose youth stands in marked contrast to the suffering husband. The genial husband depends on a self-sacrificing, loving, much younger and energetic wife to balance his mood swings and act as a catalyst for geniality: "A wife who permitted herself to cry would never have done for poor old Henry. His chronic weakness had compelled her to be unremittingly strong."[35] Huxley's Nobel Prize laureate is the foil of a negatively described genius: he is moody, egotistical and lives on another planet; in many ways, he embodies the neurodivergent scientist. Henry Maarten is completely dependent on the help and support of his family and knows how to behave in such a way that his wife would not think of being absent. Genius and madness intertwine in Henry Maarten. It is interesting that Huxley almost completely excludes the realm of science and lets his characters interact in the private sphere, the very sphere that reveals Henry Maarten's interpersonal weaknesses. The trigger for a deeper insight into the character of the Nobel Prize laureate and into the marital relationship is the forced departure of Katy Maarten, who wants to care for her terminally ill mother:

> The two children – the three children, if you counted Henry – were left in charge of Beulah and myself. Timmy gave us no trouble at all. But the others, I assure you, the others more made up for Timmy's reasonableness. (…) The Nobel Prize winner wouldn't get up in the morning, cut his lectures, was late for every appointment.[36]

Besides practical help, Maarten needs his wife as inspiration. Huxley's great genius is ultimately revealed as an old man dependent on eroticism and love, who is not viable without female companionship and so, after Katy's accidental death, he promptly marries his sister-in-law and, after her death, an even younger female scientist. The cliché of the (apparently) successful old man who always binds younger women to himself is found in its purest form in Huxley's novella. Genius is directly related to childlike, naïve delusion and sexual tension. The social shortcomings of the Nobel Prize laureate cannot be compensated for by his genius. Once again, it is the personal dimension that comes into the author's focus, for it is here that the weak points of genius reveal themselves. It seems that genius must go hand in hand with social shortcomings in literature.

## Conclusion

The analysis in this paper of several texts from different literary genres has shown how the Nobel Prize and its laureates function as representatives of societal and social power. Laureates appear as true geniuses in their respective fields, but also present the personal shortcomings of cranky or career-focused people. The diversity of functions can be justified by the respective literary tradition, but also by the context-specific characteristics. Thus, genius is used as an elusive character trait, as a foil, to drive the plot forward in the desired direction. In the thriller, we find the genius-crazy professor, just as we encounter a strongly reflective narrative instance in such development novels as *Fast genial*, which repositions and considers genius in relative terms. Genius figures, Nobel Prize laureates and the Nobel Prize itself thus become constantly rewritable surfaces that can always be rethought in their evaluation and thus imply a broad spectrum of interpretations.

The analysis has brought to light ingenious scientists who seem to be completely absorbed in their task and lose all contact with their outside world and any sense of tact. Huxley's character Henry Maarten is particularly emblematic of this oblivious but egocentric genius. We have, nonetheless, also come across well-calculated actors like Dr Lisa Michael, whose genius is rooted above all in hard work and the absence of any empathy. Heroes of science are eccentric, they are symbols of genius, power and ingenuity, but they can also fail and it is precisely this aspect that is emphasized in literature. They do not fail because of the goals they set themselves or because of their own megalomania, they fail above all in the private sphere, in interpersonal relationships, the need for love and the family situation. More drastically than any other text, Irene Dische shows this failure on a personal level through her Alzheimer's novella. Her nameless protagonist has achieved everything professionally, but has completely failed privately as a brother, husband and father and now ends up in lonely oblivion, accompanied only by the reproachful shadows of the past.

Nobel Prize texts rarely portray a perfect genius acting sacrificially. More recent literature shifts to the private sphere of geniuses, to their weaknesses and sufferings, and uses genius and honours as markers for the absence of humanity or for the unequal weighting of private and professional interests. The Nobel Prize is the highest award that scientists or authors can receive and for this very reason, the fall is enormous. The fallen heroes are ideal symbolic figures who can represent moral misconduct as well as excellence. The Nobel-biotope in (especially recent) literature is inhabited with neurodivergent figures, the solitary and socially deviant geniuses.

The novels tie the Nobel Prize to the traditional madness and genius discourses of the nineteenth and twentieth centuries. The figures shown within their working environment often embody the stereotype of the careerist. Others conform to the stereotype of (semi-)mad genius. "If Stockholm, as often suggested, is science's Heaven, it is not a paradise which is inhabited exclusively by angels and saints."[37] Thus concludes Brodesco following his study of the function of the Nobel Prize in film and television, and similar conclusions are reached in this survey-like insight into the function of the Nobel Prize as a literary motif.

## Notes

1 Alberto Brodesco, "Nobel laureates in fiction: From *La fin du monde* to *The Big Bang Theory*," *Public Understanding of Science*, 2018 (May) 27 (4), 466.

2 Nikolas Immer, "Held," in *Metzler Lexikon Literatur: Begriffe und Definitionen*, edited by Dieter Burdorf, Christoph Fasbender and Burkhard Moennighoff (3rd ed.) (Stuttgart/Weimar: J.B. Metzler, 2007), 307–308.

3 Ulrich Bröckling, *Postheroische Helden. Ein Zeitbild* (Berlin: De Gruyter, 2020), 13.

4 Thomas Zabka, "Genie," in *Metzler Lexikon Literatur: Begriffe und Definitionen*, edited by Dieter Burdorf, Christoph Fasbender and Burkhard Moennighoff (3rd ed.) (Stuttgart/Weimar: J.B. Metzler, 2007), 274.

5 Gustave Le Bon, *Psychologie des foules* (Paris: Félix Alcan, 1906), 2.

6 Le Bon, *Psychologie des foules*, 2.

7 Julia Barbara Köhne, *Geniekult in Geisteswissenschaften und Literaturen um 1900 und seine filmischen Adaptionen* (Wien/Köln/Weimar: Böhlau Verlag, 2014), 22–23.

8 Claudia Bruns, "Einige Anmerkungen zur Verbindung von Ästhetik, Politik und Geschlecht im Geniediskurs," in *Exzellenz, Brillanz, Genie: Historie und Aktualität erfolgreicher Wissensfiguren*, ed. by Julia Barbara Köhne (Berlin: Neofelis Verlag, 2020), 169.

9 Köhne, *Geniekult in Geisteswissenschaften und Literaturen*, 12–13.

10 Juliane Tranacher, *Geniekonzepte bei Daniel Kehlmann* (Würzburg: Königshausen & Neumann, 2018), 72.

11 Tranacher, *Geniekonzepte bei Daniel Kehlmann*, 88.

12 Bröckling, *Postheroische Helden*, 14, 227.

13 Steve Silberman, *Geniale Störung* (Köln: DuMont, 2016), 294.

14 Benedicht Wells, *Fast genial* (Zürich: Diogenes, 2020), 77: "Mr. Monroe said that intelligent people were in danger of dying out. While academics no longer reproduced properly, but only made a career of themselves, and geniuses in particular were often without offspring, the stupid ones were bringing one child after another into the world. He wanted to do something about that. He therefore spent vast sums of money to buy the seeds of various Nobel Prize laureates and brilliant scientists." (translated D.L.)

15 Wells, *Fast genial*, 77.

16 Wells, *Fast genial*, 77: "All the other children were rather average. And then there were even losers. With complete incomprehensibilty. You take the best ingredients and breed failures." (translated D.L.)

17 Wells, *Fast genial*, 77.

18 Andreas Eschbach, *Der Nobelpreis* (Köln: Bastei Entertainment, 2009), 85.

19 Nils Hansson and Thomas Schlich, "Performing Excellence: Nobel Nomination Networks in North America," *Notes & Records of the Royal Society* (December 15, 2021).

20 Eschbach, *Der Nobelpreis*, 85: "It has also not escaped me that most people speak disparagingly of Sofía Hernández Cruz. Because she is a woman. A Spanish woman at that – it's inconceivable that someone like that could do significant work in the field of neurophysiology, isn't it? Not to mention this unspeakable moral discussion. Such are the prejudices of our esteemed colleagues." (translated by D.L.)

21 Eschbach, *Der Nobelpreis*, 85: "This woman deserves the Nobel Prize if anyone deserves it. She thought for a few minutes and then asked exactly the one question I should have been asking myself for a fortnight but just hadn't come up with. The one question that takes apart everything I believed and puts it back together again." (translated by D.L.)

22 Carola Hilmes, *Die Femme Fatale. Ein Weiblichkeitstypus in der nachromantischen Literatur* (Stuttgart: J.B. Metzler, 1990), 30.

23 Christina Sweeney-Baird, *The End of Men* (New York: Harper Collins, 2021), 383.

24 Sweeney-Baird, *The End of Men*, 256–257.

25 Sweeney-Baird *The End of Men*, 283–284.

26 Sweeney-Baird *The End of Men*, 395–396.

27 Since the text could not be obtained in the original American version, it is necessary to resort to the German translation by Reinhard Kaiser.

28 The spelling is a conglomeration of the Yiddish form of the name and the Ukrainian spelling, whether this "mistake" is intended to mark his progressive illness or to clarify his Jewish origin remains unclear.

29 Irene Dische, *Der Doktor braucht ein Heim. Erzählung* (3rd ed.) (Frankfurt am Main: Suhrkamp, 1990), 23: "The Hudson River separated my house from their house in Drohobyc." (translated by D.L.)

30 Dische, *Der Doktor braucht ein Heim*, 23: "But the other professors respect me because I have received many prizes. One in particular, a colossally noble one that everyone wanted. Zescha was with me when it landed in my lap under very hot, very bright lights. I don't know what I did with it. I always had it in my desk. But what happened to my desk? Someone must have stolen it, burned the wood in the fireplace at home, and the medal melted." (translated by D. L.)

31 Wolfgang Iser, " Das Komische: ein Kipp-Phänomen," in *Das Komische*, edited by Wolfgang Preisendanz and Hans Blumenberg (München: Wilhelm Fink, 1976), 398–402.

32 Dische, *Der Doktor braucht ein Heim*, 23: "Everyone greets me with the greatest respect, they say 'Hello, Doctor' and some say 'Hi, Doctor.'" (translated by D.L.)

33 Dische, *Der Doktor braucht ein Heim*, 23: "This coffee house truly pays for itself for me. I have the plastic cutlery and paper plates on and use them at home. I already have enough paper plates to serve cocktail bites to all the members of the academy. The napkins say MacDonalds – funnily enough. The Dr MacDonald who won the prize for chemistry in 1942 will think they are his and I stole them from him." (translated by D.L.)

34 Aldous Huxley, *The Genius and the Goddess* (4th ed.) (New York: Bantam Books, 1963), 7.

35 Huxley, *The Genius and the Goddess*, 100.

36 Huxley, *The Genius and the Goddess*, 56.

37 Brodesco, "Nobel laureates in fiction," 467.

## Bibliography


Brodesco, Alberto. "Nobel laureates in fiction: From La fin du monde to The Big Bang Theory." *Public Understanding of Science* (May 2018), 458–470. https://doi.org/10.1177/0963662518766476.

Bröckling, Ulrich. *Postheroische Helden. Ein Zeitbild.* Berlin: De Gruyter, 2020.

Bruns, Claudia. "Einige Anmerkungen zur Verbindung von Ästhetik, Politik und Geschlecht im Geniediskurs." In *Exzellenz, Brillanz, Genie: Historie und Aktualität erfolgreicher Wissensfiguren*, edited by Julia Barbara, 161-184. Berlin: Köhne Neofelis Verlag, 2020.

Dische, Irene. *Der Doktor braucht ein Heim. Erzählung.* Frankfurt am Main: Suhrkamp, 1990.

Eschbach, Andreas. *Der Nobelpreis.* Köln: Bastei Entertainment, 2009.

Hansson, Nils and Thomas Schlich. "Performing Excellence: Nobel Nomination Networks in North America." *Notes & Records of the Royal Society* (2023). https://doi.org/10.1098/rsnr.2021.0052.

Hilmes, Carola. *Die Femme Fatale. Ein Weiblichkeitstypus in der nachromantischen Literatur.* Stuttgart: J.B. Metzler, 1990.

Huxley, Aldous. *The Genius and the Goddess.* New York: Bantam Books, 1963.

Immer, Nikolas. "Held." In *Metzler Lexikon Literatur: Begriffe und Definitionen*, edited by Dieter Burdorf, Christoph Fasbender and Burkhard Moennighoff, 307–308. Stuttgart/Weimar: J.B. Metzler, 2007.

Iser, Wolfgang. "Das Komische: ein Kipp-Phänomen." In *Das Komische*, edited by Wolfgang Preisendanz, Hans Blumenberg, 398–402. München: Wilhelm Fink, 1976.

Köhne, Julia Barbara. *Geniekult in Geisteswissenschaften und Literaturen um 1900 und seine filmischen Adaptionen.* Wien/Köln/Weimar: Böhlau Verlag, 2014.

Le Bon, Gustave. *Psychologie des foules.* Paris: Felix Alcan, 1906.

Silberman, Steve. *Geniale Störung*. Köln: DuMont, 2016.

Sweeney-Baird, Christina. *The End of Men*. New York: Harper Collins, 2021.

Tranacher, Juliane. *Geniekonzepte bei Daniel Kehlmann*. Würzburg: Königshausen & Neumann, 2018.

Wells, Benedict. *Fast genial*. Zürich: Diogenes, 2020.

Zabka, Thomas "Genie," In *Metzler Lexikon Literatur: Begriffe und Definitionen*, In *Metzler Lexikon Literatur: Begriffe und Definitionen*, edited by Dieter Burdorf, Christoph Fasbender and Burkhard Moennighoff, 274–275. Stuttgart/Weimar: J.B. Metzler, 2007.

CHAPTER FIVE

# The Post-Heroic Nobel Laureate Having Fun: A New Scientific Ideal in Post-War America

*Annelie Drakman*

**Abstract**

This chapter investigates the emergence of a new way of motivating scientific pursuits from the 1950s onwards – because the scientist enjoys it. Until this point in time, why one pursued science had primarily been motivated either by its practical usefulness or by the engendering of awe and wonder. But after World War II scientists began speaking publicly and emphatically of their own enjoyment of science. Why did it become acceptable to say that one did science for the fun of it? To answer this question, self-depictions by two of the most influential public intellectuals of the time are investigated: Richard Feynman (Nobel Laureate in Physics, 1965) and James Watson (Nobel Laureate in Physiology or Medicine, 1962). It is argued that depicting oneself as having fun was a way of displaying vitality, by means of showing off characteristics and traits such as impulsivity, charm, creativity. This was contrasted to contemporary science which was described as pompous, predictable, and routine.

**Keywords:** history of physics, Cold War, masculinity theory, the history of emotion, American 20th-century history

Among recent Nobel laureates in the sciences, it is very common to state that one is doing science because it is "fun." There are countless examples. John Kosterlitz, Nobel laureate in physics 2016 explains his career choice by saying that physics was boring but "chemistry was more fun."[1] Giorgio Parisi, Nobel laureate in physics 2021, recalls his mentor saying that "we should work on a problem only if work on that problem is fun!"[2] Carolyn Bertozzi, Nobel laureate in chemistry 2022, says that she works on carbohydrates because they are "fun" molecules.[3] Jim Peebles, Nobel laureate in physics 2019, says about his own research: "you know, it's fun."[4]

Forefronting one's personal enjoyment – doing science for the fun of it – has become one of the most taken for granted, unchallenged and obvious things a Nobel laureate in physics or chemistry can say about doing science. This is quite a recent development. Until approximately the 1950s, no scientist would have openly discussed their own motivation in this kind of language. Instead, there were two other lines of argument commonly used to explain the importance of science. Either one emphasized its practical utility (it creates useful things), or one presented the life of a scientist as devotional, producing awe and wonder, emotions which engendered reverence and respect for nature.

After World War II however, in the American context, some of the most prominent scientists – many of them Nobel laureates – reconfigured the arsenal of rationales for doing science by adding a completely new argument: "Science is fun!" Such statements, then, are indicative of a specific historical moment in the late twentieth and early twenty-first centuries. Why did it become acceptable to say that one did science for the fun of it, when only decades earlier that would have been a nonsensical, even offensive, thing to say?

This article investigates a previously unstudied but ubiquitous aspect of twentieth and twenty-first century depictions of scientific virtue: fun. It focuses primarily on the public image of Nobel laureates in physics and chemistry, as related through their own words in memoirs.

Scientific biographies and memoirs are an understudied genre in the history of science considering how widely they are read. Because of their ubiquity and wide dissemination, historian of science Thomas Söderqvist argues that scientific biography may have had the strongest cultural and political impact of any meta-scientific writing, an argument in which memoirs should be included.[5] In such texts, scientists often speak in the name of science, and it is not unreasonable to assume that scientific memoirs contribute to the recruitment and socialisation of future scientists, establish and promote scientific virtues, and shape scientists' self-understanding.[6]

This article thus contributes to an ongoing discussion about how scientific virtues and ideals have been constructed historically. These encompass not only individuals, but also milieus, organisations and fields. Scientific personae – roles or templates for how to depict individual lives – are important parts of such ideals, since trust in science still relies heavily on trust in people, and individuals are repeatedly held up as representatives for science.[7]

The theoretical framework of scientific personae helps answer questions about what kind of person, exhibiting which skills and abilities, a scientist is considered to be. Personae are collective entities, which are never fully realised at the individual level, but work as resources to draw from. Building on the definition by historians of science Lorraine Daston and Otto Sibum, I take "scientific personae" to mean templates, species, classes, or types, logically prior to the person using them.[8]

As shown in other texts in this anthology, Nobel laureates are repeatedly presented as ideals for exemplary conduct in a large variety of contexts: textbooks, websites, podcasts, museums, discussions about national identity, and so on. What Nobel laureates say about science matters. "Having fun", this most innocent, overlooked and mundane of statements, does in fact encompass a complex combination of preconceptions about free choice, volition, and engagement. It is precisely because it is considered so unimportant, a throw-away characterization – often expressed with a wink and with a whiff of reluctant confession – that "fun" is able to do so much ideological work undetected.

The article charts the historical development through which it became legitimate for adult, successful scientists to publicly state that they chose their profession and topic of inquiry because it is "fun." Although this characterization is often used to describe scientist's own lives and attitudes, the step is often short to making this a prescriptive exhortation. As one scientist stated in 2020: "If the science is not fun, it probably shouldn't be done."[9]

The goal of this text, then, is to clarify how "fun" became a legitimate way for elite scientists to describe their own engagement in science. To do this, I will describe two older scientific personae which the "fun" statement is a reaction against: the sage and the hero. I will also show early twentieth century examples of how fun was said to be external to science rather than its main motivation. I will then point out some historically contingent aspects of American masculinity ideals, making the US postwar context receptive to this somewhat controversial message. Then, I will show how physicist Richard Feynman was the perfect vehicle for this message to become widely accepted from the 1980s onwards. Finally, I will elucidate what work the idea of "fun" does in relation to science.

In a larger perspective, this is meant to give further insight into the late twenty-first century interest in what Ulrich Bröcklin has called "post-heroic heroes." This term is meant to underline that although traditionally heroic figures, built around self-sacrifice, pathos and moral superiority, have become old-fashioned and quaint in the twenty-first century, the fascination with heroism has not been exhausted.[10] Investigating scientific virtues, heroic ideals and personae can help document the societal challenges and needs to which certain types of heroes respond, which values heroes embody, what the boundaries that heroes transgress mean, and the demands heroes place upon their fans.

Fun, emphatically, is not heroic. It can certainly overlap with the same kinds of activities, which can either be seen as "fun" or "heroic" – for instance flying a bomber jet in World War II. But if one has fun while doing the activity, it is no longer heroic to do so, or at least the heroism is drastically reduced. How fun managed to subsume such lofty ideals as the scientific sage and hero, then, remains a somewhat surprising development which must be explained historically.

## "Fun" and "science" before the combination was naturalized

Before the 1980s, some scientists certainly admitted to having had fun while doing science but they did just that; admitted. They presented fun in relationship to science in three distinct ways. The first way was as something which was irrelevant to the work of life. In several narratives of scientists lives from the early twentieth century, having fun was clearly distinct from doing scientific work.

For instance, J.J. Thomson, Nobel laureate in physics 1906, described in his memoirs a situation in which he "got some fun as well as instruction from my excursions into science" by playing a prank on a friend using a microscope. To Thomson, then, fun was different from science, the latter only giving "instruction." Getting fun from science was an unexpected bonus of misusing scientific equipment. Entertainment and instruction are described as clearly separate activities.[11]

Another, more critical version of this line of thinking is presenting fun as an obstacle which must be overcome on the way towards a life as a scientist. For instance, the Dutch physicist H. B. G. Casimir wrote about the group surrounding Niels Bohr in the 1920s that he feared readers of his memoir "may gain the impression that we, young physicists at Copenhagen, spent far too much time on rather childish jokes."[12] Fun, then, is presented as a rival to science for the scientists' time. Victor Weisskopf, another physicist moving in the same circles, wrote about his years as a young student in Göttingen that "when not studying, we had lots of fun."[13] Weisskopf, crucially, describes himself as having fun when he is not studying – having fun is again something different from doing physics. In Weisskopf's depiction, "fun" – in his case working as a music critic and being a theatre actor – was something which risked side-tracking him from the scientific life. After having described his fun as "pleasant distractions," Weisskopf emphasizes that these did not manage to obscure his "main reason" for being where he was: "to prepare myself for a life in science."[14] Continuing to chase fun would have, to Weisskopf, led down the wrong path of not working hard enough. In this scenario, one becomes a scientist by realizing that fun is temporary and of little value, and instead focuses on important things, like science. Science might bring a deep sense of enjoyment and even joy, but not fun, not yet.

A second, early twentieth-century way of depicting fun as related to one's life as a scientist, was to say that fun was a means of rest and relaxation and recharging one's batteries. In this scenario, having fun is not a dangerous waste of time which needs to be given up for the sake of science. However, fun is depicted in a clearly instrumental way; fun activities are clearly separate from science, and are not their own point but can work as a support system for the truly important. Casimir expresses the same sentiment in his memoir, writing that "Scientific work, and especially theoretical work, requires intense concentration and therefore – at least temporarily – a conscious avoidance of other serious thought," which is why he and other physicists in Copenhagen "sought refuge" in "childish pranks that brought relaxation without interfering with the real work." This kind of recreation of the 'work-hard-play-hard'-variety of energy was necessary, according to Casimir, and recreation through fun was much less harmful than other methods, such as "indulgence in alcohol."[15] Fun, here, provides energy to continue to do science.

The third older way of depicting fun is saying that having fun together strengthens the sociability among scientists. This is related to presenting fun as recreation, in that fun is presented as instrumental, but here it is done with other scientists, while in the previous version it could mean doing anything, with anyone, or even alone. Science and fun here become entwined in that one is having fun with one's colleagues, often in the same spaces where science is done (the office, the lab). For instance, Victor Weisskopf describes his time working as an assistant to Werner Heisenberg as characterized by a lot of non-scientific, fun activities, like playing ping-pong.[16] To depict the fun one is having with other scientists is a way of demonstrating that one has strong, lasting relationships with important people – often, the friends one is having fun with in memoirs is especially famous people. Fun, then, is a way to show that one's relationship with an important person was strong, lasting, and not just temporary. Furthermore, fun is often used to show that the friendship was a relationship between equals. This kind of activity can be seen as an outgrowth of the kind of light-hearted playfulness which is often at the heart of student organizations for university undergraduates, where fun creates strong social bonds, around which friendships are formed among students.

It is also easy to find throwaway remarks, in the memoirs of twentieth-century physicists, to the personal enjoyment they noticed among their colleagues. Eugene Wigner, for instance, says that he "amused [him]self" by combining two different branches of physics.[17] Hans Bethe, in his memoir, mentions Enrico Fermi's "joy in the challenge"[18] of a physics problem.

There were also a few early adapters of the "science is fun" message during the first half of the twentieth century, who privately made it known that they found science itself fun. However, this was still not a message deemed legitimate to express in official memoirs during that period. The main example of this is probably Werner Heisenberg. In his 1978 memoir, Walter Elsasser wrote that he, on several occasions, heard Heisenberg say that "he liked to do physics because doing physics was 'fun'." For Elsasser, this was a great revelation, since he saw scientific research

> [...] as a matter of duty, or of personal ambition, or just to make money; or perhaps, if one had a classical education, one could think of oneself as driven by the demon of which Socrates always talks in Plato's dialogues. But to think of doing science 'for the fun of it' was a new insight into the joy of life that left me exhilarated and deeply impressed.[19]

Several other scientists said about Heisenberg he had the attitude that science was fun, and he also mentions this himself. In a 1963 interview, Heisenberg says about his student years that he found it "great fun" to learn classical mechanics, and that it was "much more fun" to solve problems himself than to only listen to lectures.[20] In his 1971 memoirs, however, he does not mention this as a motive, perhaps considering it frivolous.[21]

## Two older scientific personae: the sage and the hero

To clarify the novelty of the "science is fun!" argument, let us compare it to its predecessors. The argument that science matters because of its practical usefulness has been used for centuries and is still very popular today. It is straightforward and difficult to refute: scientific discoveries and innovations have made life easier and more pleasant for billions of people.

This argument has long been accompanied by another way of arguing for the value of science, which is that scientific inquiry will make humanity more aware of nature's wonders. These two older lines of arguing for the importance of science are strongly correlated with ideas of transcendence and self-sacrifice, and connected with two templates for presenting individual life narratives of scientists, or "scientific personae": the sage and the hero.

"The Sage" is a scientific persona centred on displaying a deep relationship to the transcendent. Before the nineteenth century, a common argument for engaging in natural philosophy was that it was a way of understanding God. Reading "Nature's Book" was an alternative to reading Scripture, a way of gaining insight into the Creator through the Creation. Scientific understanding, in this perspective, leads to insights which are important because of the feelings they engender: awe and wonder. This is believed to create respect and reverence towards God and nature, the emotional values which are the core of the sage persona. An example of this perspective is given in the memoirs of Eugene Wigner, Nobel laureate in physics 1965, who said about himself that he "revered" quantum physics, and that scientists had a duty to uncover "the full unity, beauty, and natural grandeur of the physical world."[22]

"The hero" is a related scientific persona, also aspiring to higher, more transcendent values but with a stronger emphasis on self-abasement and self-sacrifice. Around the turn of the twentieth century, it was not uncommon for scientists to describe science as important enough to warrant the ultimate sacrifice: their own lives.[23] This could sometimes be done for reasons like the sage's motivations (to learn more about transcendent values), but could also be done to assist one's fellow man, for instance by curing disease. An example from early nineteenth-century medicine is the Yellow Fever Commission, a 1900 experiment led by the U.S. army surgeon general Walter Reed to determine whether yellow fever was conveyed via mosquito bites. Several researchers tested this theory by letting themselves be bitten by infected mosquitos, a highly dangerous procedure which killed one of them. As historian of science Rebecca Hertzig has shown, the men involved in the Reed experiments "were hailed in their own time as noble sufferers, willing to sacrifice their lives for science."[24] Reed himself described his fallen comrade as having displayed "manly and fearless devotion to duty," being "oblivious of self."[25]

These older scientific ideals, the sage and the hero, pre-suppose and require solemnity. They are both built around the importance, almost holiness of science. The scientific hero persona presents the individual scientist as insignificant and expandable in relation to science. The heroism of an activity is determined by how demanding it is, and of how much of one's own comfort, health and other resources one has sacrificed. The greater the repugnance of the scientific activity, the larger the scientist's own unwillingness and its cost, the more heroic. Implicit in these arguments is a debasement of the scientist as an individual, to indicate that scientists subordinate themselves to higher ideals and disregard their own feelings. The definition of something being important includes doing it even if one does not enjoy doing it.

Thus, in both perspectives, scientists are in a sense "caught" by the importance of doing science. Their personal preferences or desires do not matter. Since they work for higher, more distant goals, their immediate, day-to-day experience of science is either insignificant or even the obstacle they heroically overcome. Generally, devotion is proven through hardships endured, and the larger the hardships, the more profound the devotion. Solemnity is a crucial aspect of both the connection to something transcendent, and to displaying the expected reverence towards it.

In this perspective, enjoying work too much would be a liability, something worth hiding and downplaying. It is hardly heroic to do what one finds fun. Furthermore, the claim that science is "fun" would be nonsensical or even offensive, presenting science as something base and self-interested. Fun would demean science, remove the connections to transcendent values and make scientists as individual persons take pride of place, whereas in older narratives, individuals were just vessels for reverence and utility.[26] In these older perspectives on science therefore, it is rare to say that science is "fun." If such a declaration is ever used – and it begins popping up in nineteenth-century science popularization – it is firmly a message directed towards scientific outsiders, almost always school children.[27] That older message, directed towards non-scientists, goes like this: "science is fun! If you begin to find science fun, you will keep doing it, and then realize its importance, which will lead you to understanding its transcendence, and to you persevering in the face of difficulty – that is, you will still pursue it while it's not being fun." That message had been around for many centuries. That the adult scientist presents "fun" as the main, or even only reason for doing science, however, is an innovation.

## Curiosity as a bridge word, legitimizing fun in science

A word which is often used together with "fun" while describing one's motivation for science in the late twentieth and early twenty-first centuries is curiosity. The two words are often used as things which go together. In a 2022 interview, Syukuro Manabe, the Nobel laureate in physics in 2021, answered the question about why he became a scientist in the following way:

> I guess that one of the things is curiosity. You are curious about something and then you think about why this is happening. [climate change] turned out to be a very fun thing to think about […]. I have been doing this driven by curiosity: why is our climate changing the way it does? I think this also helped me in my career. I'm curious about it. I was doing it not because climate change is very important for human beings. I never had that slightest idea climate was going to be so important […] What drove me was pure curiosity.[28]

Here, Manabe turns his back on the view that one should choose a scientific problem based on its importance. Instead, he is driven only by curiosity, something presented as deeply personal, something which interests *him*: "I'm curious about it".

Thus, it's a motivation which emanates from his own interests and concerns, rather than those of the world. And following his own inclinations, he is rewarded with "fun," emphasizing the link between these terms, saying that climate change is "a very fun thing to think about." Similarly, Anton Zeilinger, Nobel laureate in physics 2022, said in an interview that this was his only motivation for being a scientist: "for myself it was just, just curiosity. It was always curiosity, and still is curiosity. […] my interest is always curiosity. You know, life is short, and still I'm curious to see what will happen."[29] The most important word in this quote is "just," indicating that Zeilinger uses curiosity to emphasize the banality rather than transcendence of his engagement. He, like Manabe, is building his reasons for being a scientist on an everyday, lowbrow word, just like the word fun.

Curiosity, etymologically, stems from something worldly which attracts attention. In the early-modern period, it was often a negatively tinged word, an accusation, where following one's curiosity was seen as being distracted from more worthwhile and noble pursuits.[30] According to *Oxford English Dictionary*, two of the oldest meanings of the word are "desire to know or learn in a blameable sense, the disposition to inquire too minutely into anything; undue or inquisitive desire to learn," and "inquisitiveness in reference to trifles or matters which do not concern one."[31] Still, curiosity, unlike fun, has a long history of being associated with science, being used to justify its empirical, material concerns.

In the mid-twentieth century, science was often said to stem from curiosity. In this context, the main function of the word was refuting the argument that it should be useful. Lauding "curiosity-driven research" as the most important part of science was and is a way of defending research which is not expected to yield any immediate result apart from increased knowledge.[32] This meant praising "basic" research, intended only to win new knowledge without a use in mind. Basic science is usually contrasted with Research and Development within corporate or industrial applied research, which work on solving practical problems. Often, the defence of curiosity-driven research is built around giving previous examples of incidents when undirected curiosity led to unexpected or serendipitous discoveries. In that context, then, curiosity means taking risks, and is opposed to tinkering to achieve gradual improvements.

The main reason why "curiosity" goes well with the "science is fun" perspective is that it is distinctly secular and worldly, with a very weak link to transcendent, important topics. Rather, this trait is generally presented as mundane and idiosyncratic, strongly related to an individual's unique personality. In contrast, other terms for knowledge seeking emotions, most importantly awe and wonder, hold the same relationship to curiosity as joy does fun. Both awe and wonder stem from a long history of describing dramatic, overwhelming emotions directed towards God. Such emotions indicate being in the presence of something far greater and more important than oneself. Wonder, for instance, is said to be the beginning of wisdom by Plato, in a quote which is often said to apply to science.[33]

Curiosity, like fun, does not have any such transcendent connections historically, but rather connotes prying, sticking one's nose into other people's business. In the 1950s Albert Einstein, somewhat awkwardly, tried to amend this mundane tinge by coining "holy curiosity," but this term does not seem to have caught on. The full sentence from which this term is taken is strongly related to the "sage" persona: "One cannot help but be in awe when one contemplates the mysteries of eternity… Never lose a holy curiosity."[34] Einstein was both described as a sage in the popular press, and took on such aspects himself, speaking here of mysteries, awe, eternity, and the holy.[35] The term "holy curiosity," then, can be seen as a bridge term between the sage and the fun personae.

That bridging function mostly builds on a second meaning with which curiosity is charged in the middle of the twentieth century, where it begins to be described as the opposite of disengagement from the world. Curiosity is in that context used to describe vitality and warm-bloodedness. For instance, Richard Feynman was described by a student as her "hero" because he elevated science into a noble expression of human curiosity. The student claims that "the curiosity that led Feynman to unify quantum field theory with relativistic electrodynamics is the same curiosity that made him wander into topless bars and experiment with marijuana and sensory deprivation tanks."[36] Curiosity, then, is a human ability to deeply engage with anything. One of Einstein's most famous quotes is probably "I have no special talents. I am merely passionately curious," connecting passion to curiosity.[37]

Similarly, Eugene Wigner uses curiosity as a positive character trait, saying of Edward Teller (a physicist shunned by many for his betrayal of J. Robert Oppenheimer) that he was "intensely curious about the world," displaying a curiosity which was not disagreeable, because he did not pry.[38] Wigner goes on to depict curiosity, along with diligence and ambition, as core traits for becoming a fine scientist, more important than having imagination, a receptive mind or mental speed. Here curiosity is presented as a kind of tenacity, doggedness, and stubbornness. Because perseverance often trumps genius, Wigner argues, "even stupid people often make fine scientists."[39]

Being curious then, is presented as the opposite to being overly intelligent and flighty, quickly giving up. It also is presented as the opposite of being slow, bored, uninterested in the world, not engaged with it or wanting to understand how it works. This strong involvement with the world also explains why curiosity is a bridge word in a further sense as well, in that it connects childishness with scientific pursuits. For instance, in his 1983 memoir, Casimir quotes Otto Frisch asking "Why is it that scientists are liable to waste their time with [...] childish pranks?" and answering that scientists have "curiosity" in common with children, because "to be a good scientist you must have kept this trait of childhood [...] A scientist has to be curious like a child."[40] Here, a scientist is portrayed outright as a mix between adult and child, having kept that childish trait, curiosity, which is the core of scientific pursuits. This, then, explains why scientists sometimes waste time on childish pranks – rather than being more serious than other adults, as a result of their association with science, their scientific inclinations explain why they are more childish than others.

This, then, can explain how curiosity may have helped legitimize the "science is fun" statement. Since curiosity had been said to be the origins of science for centuries, connecting curiosity with childishness and playfulness made fun legitimate.

## The American context

There was no cultural receptiveness for the "science is fun" message before World War II, and especially not in Europe. But things were different after the war, and across the pond.

As masculinity scholar Michael Kimmel has demonstrated, the American 1830s saw the birth of a new masculinity role: the self-made man. Unlike older ideals of manliness, the self-made man is not anchored in a social context of respectability, like landownership or local roles of leadership, but is instead defined by his success in a free market.

The background of the self-made man is insignificant. Rather, his inclination to act is paramount. Within this ideal, perseverance in the face of obstacles is valued over almost all else, even ambition for success, which could be questioned as being greedy.[41] Around the second half of the nineteenth century, this ideal is combined with a tradition of manly heroism, especially as expressed through polar expeditions and other dangerous adventures. To both the self-made man and the heroic explorer, it is of crucial importance not to be dissuaded by adversity, and especially not to show any negative emotions while persevering. To be the ideal American man, in the nineteenth century, meant being emotionally cool. But once the stiff upper lip had been mastered, the ideal shifted, moving beyond emotion free to the kind of tranquil serenity and ease expressed through cheerfulness.

The message that cheerfulness was a part of the American national identity, that cheerfulness was manly, and that cheerfulness was central to not only heroism but to the kind of everyday business success the self-made man was chasing, was a message repeated in countless self-help books around the turn of the twentieth century. Central to this development was a man named Orison Swett Marden, whose books achieved unprecedented success, being printed in some 250 editions.[42]

In the nineteenth century, the self-made man persona had not been perceived as cheerful, and it took a great deal of work to bring these two ideals together, but it solved several problems. Cheerfulness is a social lubricant, overlapping with friendliness and trust. It establishes a shallow, temporary sociability, which promotes quickly established but brittle social work relationships. This attitude makes business easier by improving relations with customers, employees, collaborators, and partners. It was said to increase the self-made man's self-composure by creating resilience.[43] Most importantly, cheerfulness can provide a strong defence against capitalist vices, like being driven by greed. Avarice is almost always described as a gloomy, joyless mood: being "cheerfully greedy" is a contradiction in terms. A cheerful attitude, then, through its connotations to generosity, can make criticism of capitalist ruthlessness less clear and effective. Through cheerfulness becoming a core aspect of a highly valued masculine ideal, the ground was prepared for grown men to start talking about having fun, a change from earlier masculinity ideals.[44]

Thus, since the early twentieth century, the self-made man existed as a ready persona for American men, combining ideals of being hard working with being cheerful.

There is also, as mentioned above, a long Christian tradition of lauding outwardly expressed positive emotion, especially joy. In Protestant circles, in particular, joy is often presented as a theological aim, expectation and reward. In the New Testament, believers are repeatedly encouraged to "rejoice," and many protestant congregations have taken it upon themselves to make sure their followers do so. Claiming that their followers were especially joyous also became a core feature of several new American churches emerging in the nineteenth century, such as the Church of Latter-Day Saints (Mormons), and Jehova's Witnesses.[45]

Thus, to say that one feels joy in one's relationship to the activity in which one engages has been unproblematic for centuries and in many cases both desired and even expected. Joy and cheerfulness, as emotional stances, also bleed into the emotions at the core of making a profession into a calling or a vocation. All this taken together meant that the American context was uniquely prepared for accepting the message that "science is fun" as true.

The twenty-first century comfort with talking about things one enjoys with the word "fun" is not unique to science, but a part of a larger change. According to Google Books Ngram Viewer, which analyzes how common English words are in their digitized collection, "fun" shoots up like a rocket – a real hockey stick curve – from the 1980s onwards. Between 1980 and 2013, the frequency of use of the word fun almost quadruples, having remained almost unchanged 1820–1980.[46] It is also used in fields other than science – for instance, the memoir of the creator of the free programming language Linux gave his book the title *Just for Fun*.[47]

Furthermore, after World War II there was a widespread disillusion with charismatic leadership and self-effacement in response to higher ideals. As historian Ulrich Bröcklin points out, "the Nazis' unparalleled crimes against humanity were possible largely due to their mobilization of a militant heroism that preached fighting unconditionally unto death as heroic self-sacrifice."[48] In the postwar period, self-sacrifice, which had been at the core of the "scientific hero"-persona, thus went out of fashion. The "scientific sage" was also increasingly seen as a relic from an older age. While suitable for neo-gothic university palaces with monastic histories, the sage was an anomaly in modern office buildings. Besides, the two most influential American scientists who embraced the sage persona disappeared from public view in the mid-1950s: In 1955, Oppenheimer's security clearance was revoked, and in 1955 Einstein died.

The sage persona did not completely disappear. It is sometimes still used within some specific fields of science, primarily mathematics and astrophysics. But after the 1950s, it was marginalized. Another reason for why it, along with scientific heroism, was eclipsed is that ideals relying on inner conviction were not very useful during the Cold War climate. At the core of both older scientific personae, the sage and the hero, was being self-motivated, without having to be convinced of the worth of science. But in the Cold War context, once the launch of Sputnik in 1957 supposedly demonstrated the technological superiority of the Russians, many Americans suddenly saw it as crucial to make science seem attractive to the masses. As thousands of new scientists and engineers were needed to win the Cold War, science popularization gained an unprecedented geopolitical importance. In this context, loudly repeating that "science is fun" became a reasonable thing to do. One of many examples is a 1958 text in *Scientific American*, quoting physicist Edward Teller saying that although scientists were crucial to overcoming communists, "a scientific career is more than a duty, though. It is an opportunity. Science is fun."[49]

This message also coincides with a great deal of critique against science from the mid-1960s onwards from feminists, environmentalists and pacifists. Science faced critique both for its historical practices, like excluding women and minorities from their ranks, current practices, being the means for their subordinations through classifying them as deviance from the norm, and finally for threatening to bring about a horrifying future. To scientists, suddenly responsible for bringing the world to the brink of nuclear apocalypse, the "fun" message was a welcome new perspective.

"Fun" also goes well with a new kind of discourse in the American 1960s, preoccupied with the usefulness of creativity and play. In his most famous book *The Farther Reaches of Human Nature*, Abraham Maslow, probably the most influential American psychologist of the 1960s, wrote that "the puzzle I'm now trying to unravel is suggested by the observation that the creative person, in his inspirational phase of the creative furor, loses his past and his future and lives only in the moment. He is all there, totally immersed, fascinated and absorbed in the present, in the current situation, in the here–now, with the matter-in-hand."[50] Maslow does not use the word, instead he mainly talks about "peak experiences" but this sounds awfully similar to having fun. Maslow, promoting what he called "self-actualization," argued that individuals following their own, most personal inclinations and volitions, became the most personally harmonious and productive of all. This message appealed both to organizational theorists, trying to increase productivity within the business sphere of corporate America, and to counter culture. Depending on the context, Maslow was used to legitimize attempts to increase worker efficiency, or to turn one's back on wage labor altogether. The most lasting influence of Maslow was probably the way he legitimized a strong interest in authenticity. In relation to joy, cheerfulness, and fun, this would promote fun rather than cheerfulness. Cheer was inherently shallow and vulnerable to accusations of insincerity, since cheerfulness is generally meant for others and is an emotional self-presentation (the word cheer originally only meant "face," or "facial expression") put on regardless of felt emotion.[51] Fun, however, combines the authenticity of felt joy, which was an emotion which could be privately felt and was not necessarily expressed socially, with the outgoing gregariousness of cheerfulness.

## The main proponent for "science is fun": Richard Feynman

Richard Feynman (1918–1988) is perhaps, after Albert Einstein and J. Robert Oppenheimer, the most famous American physicist of all time. He was awarded the Nobel prize in physics in 1965 for his work on quantum electrodynamics, but, to the general public, he was even more famous for making science seem *fun*.

Richard Feynman first became known outside of physicists' circles through a 1983 BBC2 television series about science called *Fun to Imagine*, named after a Feynman quote.[52] To the majority of the American public however, he became famous for participating in the 1986 NASA inquiry into the Challenger disaster, effectively demonstrating a crucial design flaw by dropping a piece of rubber into ice water during a televised hearing.[53]

In the mid-1980s he released two memoirs, *Surely You're Joking, Mister Feynman: Adventures of a Curious Character* (1984) and *What Do You Care What Other People Think? Further Adventures of a Curious Character* (1988).[54] Both were collections of anecdotes which a collaborator had recorded during drumming sessions. Since the memoirs are based on transcribed interviews they have a distinctively casual tone, as if the reader is being told a story. The books became classics and have remained on best seller lists of scientific memoirs for decades.[55]

Feynman became famous not only for his scientific insights but for being a fun person, having unconventional hobbies: playing the bongo drums, cracking safes at Los Alamos during the Manhattan project, and picking up women. When published, *Surely You're Joking* was received as pushing the boundaries of the genre. Reviewers noted its unconventionality but across the board, they were charmed: *The Washington Post* reviewer calling his book "delightful."[56] *The New York Times'* reviewer wrote that many science buffs were "going to be unnerved" by Feynman's self-display, sticking "out his tongue at some of our most cherished scientific institutions." Yet, he admits that "in the end, one winds up not only forgiving him, but admiring him."[57]

Although many parts of his memoirs concern life outside of work, at the core, his message is that science is fun. In an often-quoted anecdote, Feynman is a burned-out new professor after World War II, describing how he used to feel:

> I used to *enjoy* doing physics. Why did I enjoy it. I used to *play* with it. I used to do whatever I felt like doing – it didn't have to do with whether it was important [...] but whether it was interesting and amusing for me to play with [...]. I'd invent things and play with things for my own entertainment. So I got this new attitude. Now that I *am* burned out and I'll never accomplish anything [...]. I'm going to *play* with physics, whenever I want to, without worrying about any importance whatsoever.[58]

When Feynman put this new playful attitude to work, he made the discovery which led to his Nobel Prize. Feynman showed his work to a fellow physicist, who asked why he was working on it. Feynman answered "'Hah! There's no importance whatsoever. I'm just doing it for the fun of it'."[59] This new attitude of letting fun lead the way is said to have saved Feynman from burn-out, making work "effortless. It was easy to play with these things. It was like uncorking a bottle: Everything flowed out effortlessly. I almost tried to resist it!"[60] Fun, here, is described as a way both of resisting the temptations of conforming to other people's expectations (by doing work they deem important) and as a way of rediscovering one's energy and enthusiasm, necessary for the long hours needed to make important breakthroughs.

It matters that it was Richard Feynman who became the most outspoken informal spokesperson for the importance of having fun in science, and not someone else. As previously shown, he was certainly not the first scientist to talk about finding science fun, but I would argue that he was the one who made this into an unconventional, safe claim to make, preparing the grounds for its ubiquity in the twenty-first century.

Feynman was the ideal spokesperson for presenting science as fun since this was still a controversial message. Fun, in the 1980s, retained connotations to people with low status, like children as well as to laziness, leisure and rebellion. Despite the cultural groundwork having been done for connecting elite scientists to informal fun, this would have been a difficult sell for anyone except the most admired kind of person.

But Feynman was exactly that in the 1980s – admired and celebrated. And then, just two years after his Challenger Inquiry triumph had made him a household name, he died from cancer. Other well-known Nobel laureates in science, such as Linus Pauling or Jim Watson, lived for decades after having become famous, giving them plenty of time to tarnish their own reputations. But Feynman became a household name through his televised success, and then promptly died. His memoirs filled the void he himself left behind, and they quickly became classics, often given as gifts to undergraduates in science even today. Apart from mild controversy in the 2010s regarding whether his attitude towards picking up women displays sexism and misogyny (as an article, entitled "Surely You're a Creep, Mister Feynman," claims), Feynman remains beloved in almost all quarters of American society, a unique position for any scientist to have.[61]

For instance, in a 2020 review of a biography of Feynman, the reviewer writes in *Isis*, the most revered history of science journal that

> Feynman was a physicist easy to admire. He continues to exert wide fascination and is idolized by many a young physics student. He is admired not only for his outstanding achievements in modern physics and his genius but also for his reckless personality, his charisma, and the fact that he visited strip clubs and played the bongos. He was one of a kind, as handsome as a Hollywood actor, and he loved to entertain. There was a sense of exuberant youthful impulsiveness about him. Anecdotes about him fill books, and Feynman himself published numerous popular works. His brisk, firm, and witty statements are often quoted. For these reasons Feynman's fame reaches far beyond the narrow physics community.[62]

This is hardly a lone voice. There are many other examples. A student, writing an assignment about Richard Feynman as "my hero," described him as being "a great teacher endowed with a hopelessly attractive and idiosyncratic character," the antithesis of the stereotypical physicist who was "somber, subdued, and anti-social."[63] What was it about Feynman that made him so attractive? The student's answer is that it

> was his fun-loving and child-like curiosity with nature, disdain for pretension and hypocrisy that charms the aspiring physicist into admiration. Feynman makes physics fun. And before I even enrolled in my first rigorous math or physics class I had read mostly all of his popular physics books, each one reinforcing my own innate sense that the study of nature is, at its core, a playful exercise in which everyone, lay or expert, can participate and understand.[64]

Feynman was the perfect vehicle for introducing "fun in science" as a motivation for becoming a scientist because in several aspects, he was "the ideal man" of the 1980s. No one else had the combination of social capital – scientific, authentic, and charismatic – which he held at the time his memoirs were published.

First, his scientific social capital was almost impossible to match, having been a professor at Caltech for decades, and being a Nobel laureate who remained a highly respected scientist among his colleagues until he died. In his 1988 obituary, the president of the *American Physical Society* described him as "the most creative theoretical physicist of his time and a true genius," and fellow Nobel laureate Hans Bethe called him a "magician."[65] The respect from one's peers, unusual in Feynman's case in that it remained almost unanimous, is highly useful for becoming a scientific role model.[66] Also, being a member of the commission explaining the Challenger disaster – a task force ordained by president Ronald Reagan – demonstrated publicly that he was well connected to the techno-scientific and political elite.

Second, he came across as authentic. Feynman spoke with a New York accent strong enough for his peers to suspect exaggeration, two colleagues even suggesting that Feynman spoke like a "bum."[67] His crude and unpolished manners were often remarked upon, the physicist Freeman Dyson called Feynman "half genius and half buffoon."[68] This was a virtue in disguise. It made him seem *real* during a period when unabashedly flaunting one's origins had become a virtue more highly valued than adapting to new environments. After decades of Cold War anxieties about spies, people hiding their true selves, this made him seem like he was proud of where he came from, rather than putting on insincere airs. Many comparisons have been made between the self-presentations of Richard Feynman and Murray Gell-Mann, fellow Nobel laureate, sometimes collaborator and Feynman's colleague at Caltech. While comparing these two men who both had working-class backgrounds, Gell-Mann, proud of being a polyglot interested in literature, is often portrayed as fake and pompous.[69] Feynman, by contrast, was seen as everyman, and proud of it. While solving the Challenger riddle, this made Feynman come across as an "underdog" both by virtue of his working-class accent, and by using simple language.

Third, he was approachable. Feynman was good at not being overly didactic, talking down to his audience but rather treating his audience as peers. His use of simple language made him easy to understand by large audiences. Also, the sheer number of people who could identify with a person who spoke with a heavy accent and had a twinkle in his eye is larger than the ones who recognize quotes from the Bhagavad-Gita. The biggest action heroes of the 1980s, Arnold Schwarzenegger and Bruce Willis, both combined an over-the-top masculinity bravado with an underdog attitude (the German accent for Schwarzenegger, the simpleton persona of Willis), along with a distinct humor meant to poke fun at pomposity. Being open about one's origins was a virtue of the moment.

Fourth, Feynman was clearly charismatic. In his 60s in the 1980s, he remained handsome, retaining a full head of hair and a slim physique. While reading his memoirs, the readers learned that he was an excellent story teller with an air of tragedy, having lost his first wife to tuberculosis in his 20s. Many of his stories also revolve around being attractive to others, especially women.

At the core of his success was displaying an ideal kind of American manliness, as evidenced by his sexual conquests and his down-to-earth working-class truthfulness.[70]

In fact, he can be said to be the opposite of that caricature of intellectuals which was so predominant in American Cold War popular discourse, "the egghead." The egghead was overly intelligent, often a scientist, who was displayed as being emasculated, unattractive, bald, fat and weak, left-leaning, effeminate and possibly homosexual.[71] Feynman, by contrast, was clearly heterosexual and very successful at picking up women before settling down into middle-class conventionality, getting married and having two children.

In conclusion, in the 1980s and onwards, Feynman became close to an ideal scientist. He was a member of the scientific elite, combining decades of being highly successful and respected within the scientific community with an air of recent success after the Challenger Inquiry. He was an attractive man, displaying masculinity ideals which corresponded well with 1980s conceptualization of heroism in American popular culture. As opposed to other famous physicists, like Oppenheimer or Einstein, Feynman came across as uncomplicated with no signs of inner angst, melancholia, or *Weltschmertz*. He also managed to appear like an underdog by proudly displaying his working-class background, which helped him come across as an authentic, direct truthsayer, not swayed by polite society or ensnared by bureaucracy or careerism, but rather calling a spade a spade. He was also immune to the popular taunts of intellectual men in Cold War America, especially being called an egghead, because of his charisma and heterosexuality and being uncommonly successful with picking up women.

Feynman's image, then, was resilient enough to be able to present a slightly controversial statement as unabashedly true: that science was fun. This message, in fact, did not destabilize him but further emphasized several aspects of his self-depictions: that he was not afraid to tell the truth, that he was driven by the love of science rather than by careerism, and that he was playful and therefore creative and open-minded.

## What does fun do?

In conclusion, from about the 1980s onwards, an increasing number of scientists began using the word "fun" to describe what science is like. But what is the innovation here? What is new?

It's not the positive, emotive aspect. That has been expressed through words like "joy" and "love" for many years. Two examples are Victor Weisskopf, professor of physics at MIT, who named his memoir *The Joy of Insight,* and Eugene Wigner, 1963 Nobel laureate in physics.[72] Wigner says that "it is a joyful thing to know that you are truly a physicist. What else besides love can compare with it?"[73] Wigner is likening the joy he feels for physics to love, which means presenting joy as a strong, long-lasting bond, the kind of connection a person shapes their life around. There are several similarities between talking about one's joy and love for science and saying that science is fun, primarily the positive evaluation of the fun or joyous activity. Both words are used to say that the person enjoys it, wants more of it, and wants to keep doing it.

Because of these similarities, "science is fun" has not previously attracted any attention as surprising or noteworthy. But this *is* an innovation, because fun introduces several new connotations which were not present in the older configuration between science and joy.

## Differences between "joy" and "fun"

Fun denotes the mundane – joy denotes the important. The term joy has old roots in centuries-long discussions about moral virtue. In Christian theological discussion, especially in Protestant circles after the reformation, joy was said to be both generated by a virtuous life, and proof of God's grace, a sign of coming salvation. Felt and expressed joy is therefore presented, for instance in Martin Luther's writings, as a sign of being a true believer.[74] In Christian discourse, joy is also often said to be the ultimate reward. For instance, in sixteenth-century English sermons worldly joys were presented as precursors of the eternal, divine joy one could be rewarded with after death.[75] Thus, in the twentieth century, the word joy has a deep resonance, subtly hinting at historical connotations to the good, ethical life, living well, and being justly rewarded.

Other terms for personal enjoyment have similar backgrounds, having been used to define religious emotion. Especially the words "passionate" and "enthusiastic," often used about science, hint at the older, theological meanings of the words.[76] These words, like joy, have long histories of denoting a strong bond between the individual and the object of the feelings – an emotional connection so powerful it might threaten to overtake, even enslave, the individual's free will, hinting at self-sacrifice. The word "passion" refers to both the days when Jesus was crucified to atone for the sins of mankind, and strong, pleasurable engagement in something. This kind of language is compatible with the older "hero" and "sage" personae because it goes well with self-sacrifice and self-effacement. But while joy, passion and enthusiasm all have long traditions of describing the emotions believers have felt in response to the divine and the transcendent, fun has no such roots.

Fun indicates a milder engagement. That in itself is not unique – in twentieth and twenty-first century usage, words such as passionate, enthusiastic, and joy have been secularized, and their intensity much reduced since previous centuries. They also describe temperate enjoyment, no longer all-consuming. Thus, in contemporary use, they can be synonyms to fun.

But the reason why it becomes possible to say that "I find science fun" and take it to mean approximately the same thing as "I love science" or "I feel passionate and enthusiastic about science" after World War II, whereas previously such statements would have been offensive or nonsensical, is because "fun" introduces several connotations which are not present in the other words.

Joy, enthusiasm, passion and fun all have enjoyment in common, but fun stands out in several ways: it is fickle and short-lived, and centered around an individual's will, forefronting freedom. It also has a distinct everyday character. Fun has no relationship with the transcendent. No higher values are indicated through fun. Fun is here and now and nothing more. Having fun is the opposite of attitudes required and expected in relation to the transcendent like solemnity, seriousness, and reverence.

Etymologically, the English word "fun" comes from fooling someone, doing pranks, telling jokes.[77] It describes worldly activities without connotations to higher spheres of divinity, transcendence, morality, or virtue. Fun is clearly a positive evaluation, but carries no hints of what is fun is somehow "right," "fair," or "just."

A marked difference between personal engagement which presupposes transcendent values – like joy and wonder – and activities called fun, is that the latter can be still be fun even if they are meaningless. While joy and wonder are tied up with something being important and meaningful, that is not the case for fun. A fun activity's lack of meaning is one of its main characteristics, since the outcomes of a fun activity by definition are unimportant or secondary. Something done for fun is done for its own sake, not to reach a goal but as an end in itself. Thus, a fun activity does not need to have any meaning or importance, no connection to a higher goal or aim.

Fun, however, has that in common with the transcendent that it elevates everyday life. Having fun while doing science is often likened to seeing science as an adventure. This would be an adventure without risk or cost, available to anyone, all the time, through only the right mindset. Seeing scientific work as a fun adventure means having safe, easy, immediately accessible adventures without antagonists, not requiring travel, cost or any kind of hardship. It transforms humdrum, boring environments, for instance bland industrial labs, into something exciting.

From the 1960s onwards, there is a shift in meaning, where having fun begins to be described as meaningful in a different sense – not as a way to worship a transcendent, higher value like nature or God, but meaningful for the individual experiencing it as a kind of self-exploration.

This perspective is expressed by proponents of humanistic psychology, a group of American psychologists working from the 1960s onwards to identify the origins of human motivation. It became most firmly expanded upon by the psychologist Mihaly Csikszentmihalyi through his term *flow*. Coined in the 1970s, the term means strong, pleasurable engagement in work, which leads one to lose track of time – often likened to having fun.[78] Fun, then, becomes fully legitimate to the working life – in fact, the most legitimate way of engaging with one's work.

It is also a way of charging science with *personal* meaning, relating to the individual, creating meaningfulness in the sense of exploring one's own preferences, without resorting to references to higher, transcendent values.

This ties into what Steven Shapin calls "the moral equivalence of the scientist," a strong line of argument in the postwar period, where many scientists repeatedly stated that the universe is meaningless, and that scientists are not moral authorities since science can only help explain what exists, not what ought to be done.[79] To say that science is "fun" in that context is to say that science is meaningless, in that it is not transcendent. Scientific research will only give insight into that which is being investigated, not into higher structures beyond, such as God's plan.

But simultaneously, the positive evaluation embedded in "fun" makes science meaningful and important to the individual making the statement. "Fun" puts the individual at the center of making science meaningful instead of God or nature.

Generally, joy is described as a reactive emotion to something external, which triggers joy. Fun, however, is generally taken to emanate from the individual. It would be wrong not to take joy from a joyous event, but not being amused by an event meant to be fun is a way to distinguish individuality. The individual is always in charge in determining whether something is fun or not for them. That is, fun is something individuals can claim themselves without external anointment.

This is also different from the positions of being a sage or a hero, which must be ascribed by others. They are not designations one can claim for oneself, but are constructed by others' admiration. Calling oneself a hero would negate the purpose, since this virtue includes selflessness, while aspiring to a label as honourable as "hero" is obviously self-serving. Thus, the two older personae require the recognition of others. "Having fun," however, does not; only individuals themselves can determine what they find fun. Furthermore, fun does not bind the individual to the activity in the way that joy or passion can do.

Fun is shallow, temporary and fugitive. Having fun is often described as the opposite of grit, determination and hard work. Since fun has no relationship to transcendent values, a fun activity can only be expected to be pursued for as long as it remains fun, and can then, legitimately, be abruptly abandoned. There is no point in persevering through boredom or frustration if the only point of an activity is its fun. Perseverance only makes sense if there is a disconnect between the current experience of the activity (it's hard, I want to quit) and a future reward (this will be worth it in the end). A fun activity, however, has no "end" in sight, being its own reward. This puts the individual having fun firmly in control. Fun, thus, indicates weak ties to science as an activity. Scientists calling science fun are implying that science may be abandoned as soon as it is not fun anymore, which makes this kind of attachment seem loose and non-binding; there are no chains of devotion or reverence to trap the scientist.

But here is a paradox. Calling science fun indicates that one could abandon science at will. Not having done so, but having stayed, proves one's willing devotion to it more strongly than saying that one felt compelled to do science because of its importance. Calling science fun is usually done in retrospect, in interviews or memoirs after decades of hard and successful work. In that context, the lightness of the "fun" kind of attachment becomes surprising. Nobel laureates' messages, that they have been doing science because of the fun of it, is impactful because it contrasts with the importance of the work performed, which is what was being acknowledged by a Nobel prize. "Fun" shows that the fun-loving scientists have not acted the way fun would have permitted them to act; they did not give up at the first opportunity, but rather kept at it long enough to become successful.

"Fun" thus has similarities to how "joy" was used in Reformation Protestant groups to distinguish true from false believers. A true scientist – someone like

Richard Feynman – is never bored by science, but constantly finds it fun. This is demonstrated by the fact that since he is not bound to science by its importance, he could have left at any time. That he did not, but instead voluntarily worked hard for many decades, proves his devotion.

That is a very high bar to live up to. No longer having fun would here mean a loss of devotion. It is no longer enough to be good at science, and to realize that it is important, and thus persevere. Rather, one must constantly feel the right kind of strong positive emotion. Such demands on one's emotional state presumably leads to self-regulating management techniques meant to control one's own moods, volition, and desire. Foucault called such management "technologies of the self."[80] As soon as the bond of strong emotional engagement and enjoyment wavers – when one feels bored, sick of it, tired, wants to quit, interested in something else – one has stopped being the kind of person who always finds science fun, and thus stopped being a "true believer" of science.

Refusing to find science fun, then, would be risky in twenty-first century discourse. It would imply that one is not a true believer but rather working as a scientist because of illegitimate desires for such external rewards as fame or career success, or that one is reverential towards science, subscribing to the "sage" or the "hero" personae, both of which have gone decidedly out of style.

Having fun does have something in common with the old "sage" and "hero"-personae: that it forefronts being self-motivated. Since one enjoys the fun activity, one always wants more of it, and is in principle uninterested in any compensation such as salary and recognition. In this disregard for rewards, fun is especially similar to the older ideal of scientific heroism. The refusal of compensation was a crucial component of constructing someone as a hero. In describing the heroism of his 1900 Yellow Fever commission, Reed emphasized that the men who purposely let themselves be bitten by mosquitos which they suspected spread yellow fever "accepted the life-threatening task only on the condition that they receive no financial compensation for it."[81]

The emergence of fun as a value has political implications. In contrast to joy, which implies devotion and transcendence, fun severs the link between boring minutiae and higher values. Fun makes transcendence pompous, and makes sacrificing oneself for a higher ideal seem like tiresome martyrdom. This means political systems of change which depend on people subordinating themselves to a larger, more important value, for instance nature, will lose ground. Political, unified action becomes difficult. Of course, the unraveling of the social fabric – the fading away of civil, voluntary society, is not accomplished simply by "fun" as a value. This results from a complex situation with several reasons. But fun, being easy to understand and irrefutable, is often the shape resistance takes: "why should I go to a political meeting? Where is the fun in that?"

A clear difference towards older ideals, then, is that fun is amoral, unrelated to doing good for others. Having fun is often presented as a "free zone" from morality and outside concerns. Jokes are normally funnier the stronger they violate boundaries which restrict polite society: generally, tolerance of crudity or cruelty is larger while joking, because joking hinges upon the unexpected.

Doing science because it is fun rather than because it is important frees the scientist up to do science irreverently, sloppily, backwards, any which way. While having fun, everything is possible. Rules and restrictions are meant to be toyed with or ignored; anything goes. Fun reduces the need to show reverence towards the scientific object one is trying to understand, one's equipment, or the arenas where science is conducted; the high status of science is reduced. The lab becomes a bouncy castle, not a church. Such a view goes very well with creativity, a concept which increasingly has become applied to science from the 1960s onwards.

## Connections to creativity

Creativity, a concept meaning the ability to generate new and useful things, is an American twentieth-century invention. In Cold War America, from the 1950s, cognitive psychologists worked hard at changing conceptions of what lies at the heart of original creation, abandoning the nineteenth-century preoccupation with "creative imagination" and replacing it with "creativity." By presenting "creativity" as something which could be assessed and nurtured, and then spreading this view in popular media, they managed to reform conceptions of creative ability from a mysterious insight and ability only present in a small elite of exceptional geniuses, into something mundane. Creativity, in their view, was a potential in every human, and was associated with small, insignificant changes just as much as with masterful breakthroughs.

"Creativity" was presented as a human trait worth psychological study in a 1950 speech by J. P. Guilford, the president of the *American Psychological Association.* This ability was very much of the period, and seen as a fruitful way forward between the extremes of unimaginative conformity and destructive genius cult. Creativity was repeatedly connected to non-conformity and playfulness, but also seen as strongly related to productivity, and was used to identify the most promising future students and workers through a battery of tests. In this way of thinking about original creation, fun was seen as an aspect of playfulness, which was seen as an aspect of creativity, which was seen as a crucial part of productivity. Thus, fun became not the opposite of work, but a sign of highest possible engagement and future useful output.[82]

## Childlike innocence

To combine fun and science means associating science with groups of much lower status, primarily children. Several contemporary scientists who often call science fun openly make this connection, for instance likening scientific equipment to "cool toys in the lab,"[83] as Donna Strickland did in her 2018 Nobel Banquet speech, or like John Clauser, saying that he as a child, visiting labs, thought that "I want to be a scientist so that I can play with neat toys like this."[84]

The connection between children and fun is long established and unproblematic. Combining being childish and doing science is not new either, but this was a very different trope before the 1980s. Generally, if scientists likened themselves to children, it was something done in line with the sage and hero personae, because they were doing it to put themselves down, emphasizing their own insignificance and ignorance. This, then, is a way of showing the importance of science by making oneself seem small in relation. The most famous example of this is perhaps Isaac Newton's claim to only having been a child collecting pretty rocks at the beach of the vast sea of nature.[85] Childishness in relation to science has also been used, although more rarely, to indicate innocence and purity, to say that one's intentions have been good, that one has a proper sense of the awe and wonder of science, and that one is not debased by any such presumably adult preoccupations as greed, desire for fame, avarice, or cunning ambition, which would be base and degrading.

Emphasizing childishness in relation to the "science is fun"-ideal is different, however. Rather than indicating innocence and purity, it forefronts a lack of reverence and accountability. Saying that one sees one's scientific work as being a child, playing with cool toys in the lab, is again underlying how comfortable one is with this world, how familiar and easy scientific work is.

Furthermore, saying that science is fun is a good way of making the scientist approachable to the general public. Everyone knows what fun is and wants to have more fun in their own lives. Calling science "fun" shows the scientist being driven by similar rewards as everybody else. Thus, making "fun" a central part of the scientific persona means distancing oneself from the idea of the scientific genius, who has connotations with the Romantic artist being chosen by God as a vessel for divine inspiration. It also means distancing oneself from the "scientific sage" persona, especially aspects often ascribed to it such as having an aristocratic or upper-middle class background, being a well-read, well-traveled polyglot dripping *Bildnung* from their fingertips: think of figures like J. Robert Oppenheimer, but also scientists escaping Europe before World War II, such as Eugene Wigner.

The message that scientists find science fun emerged during the mid-twentieth century, and this should be put in relation to the growth of science popularization as a geopolitical tool during the Cold War. Having fun thus makes the scientist himself understandable, normal, not deviating from the norm in his interests but rather being a person very much like everyone else. Fun also dismantles danger – the fun scientist can counter a negative stereotype of the scientist as threatening, such as Dr Strangelove, or caricatures of Robert Oppenheimer.

During the 1950s and 1960s, scientists came under attack from several different fronts which questioned them as unattractive and pompous, aloof, life-denying, and worst of all dangerous. This critique came in two main forms. First, that scientists are boring, pompous administrators with no life: dry, polite, slow, and stuck in the past. A related critique was that scientists are unattractive "eggheads" who stink, dress badly, and are romantically off-putting.[86] Second, that they are out of control and dangerous, seduced by their own hunger for power, which might lead to nuclear war or genetically altered species running amok. The dangerous scientist is very much a contemporary theme in the post war period. The film *Doctor Strangelove: or how I stopped worrying and began to love the bomb*, Stanley Kubrick's black comedy about an insane scientist starting a nuclear war, was released in 1964, putting the finger on lived fears among a large part of the population.

Saying that science is fun can also be a way of opening the doors of science to the general public. Finding science "fun" is a mental state, which is not, in principle, dependent on being a specific kind of person: whoever expresses enjoyment, enthusiasm, excitement in relation to science is, in this perspective, a true believer, even if they do not map onto the traditional background of the scientist. Thus, it becomes extra important for women, people of color, and for other individuals who do not conform to the racial and gendered norms of science to strongly express their own commitment to science by forefronting and underlining their own passion for and enjoyment of science: "fun" becomes a legitimizing argument, showing that someone who does not "look the part" can still be a true scientist. This aspect of fun, that anyone is welcome who expresses the right attitude and level of enthusiasm, is, I would argue a result of the input from the self-made man persona, in combination with ideals about the person ideal for realizing *The American Dream.* Of course, this accessibility and equality is more of a hope than an actuality in many circumstances. It is also devious in that it makes structural challenges, like gender imbalance in science caused by sexism and racism, into failings of the individual's emotional engagement.

Still, I would argue that this message is very popular because it is seen as empowering. The persona built around doing science because it is "fun" builds on the self-made man persona in that it puts the task of choosing whether one is cut out for science in the hands of the aspirant, rather than that of gate keepers. If one finds science "fun," then being refused entry to something one desires (being accepted into a university programme, getting a job, winning a grant) is just an obstacle which needs to be heroically overcome, rather than a true evaluation of one's suitability. "Fun," by making *zeal* central to being a real scientist rather than ability, talent or skill, encourages aspiring scientists to continue in the face of adversity.

## Conclusion

After World War II, the two scientific personas the Sage and the Hero are marginalized in science, which leaves a vacuum, a need for another way to motivate science to emerge. The old argument about the practical utility of science is in fact a perennial bestseller, especially to gain funding. But it has several downsides. It is base, materialistic and worst of all un-inspirational, taking all the elevation out of science, making it sound dangerously similar to gradual improvement.

Enter, then, a new way of talking about science, which remains firmly rooted in the everyday world, clearly understandable to everyday people. Doing science because you find it fun severs you from responsibility, from connection to transcendence and to higher values. And, at the same time it is wildly attractive, mainly because it makes engagement seem free and dependent on the individual's will. While you are having fun, you can walk away any time you want. Ironically, this kind of engagement is exerting its own kind of binding force, because scientists not finding science fun can be suspected of having such illegitimate ulterior motives as careerism. Having fun in science began to be tentatively expressed in memoirs and interviews produced in the United States after World War II, while scientists previously had presented fun as either something completely separate from science or something which only restored lost energy and strengthened social bonds.

The turning point, I argue, was the 1985 publication of Richard Feynman's *Surely You're Joking, Mister Feynman.* Feynman, who held a kind of royal flush of scientific, public and charismatic capital, became a celebrity through the televised *Challenge Inquiry* and then died just as he had become a household name. His strong position made it possible for the message he was promoting in his memoirs – that one should do science only for the fun it gives, nothing more – to turn into received wisdom, where it previously had been a slightly risky message because of its associations with children, laziness, and nonworking groups. Having fun in science became legitimate.

When "fun" was introduced as a third reason for doing science after reverence and usefulness, it meant combining spheres previously depicted as opposites (leisure vs work, childish play vs adult responsibility), and moving a message previously meant for children or novices to the *core* of doing science. This innovation has profoundly shaped how science is depicted today. Ever since, the number of American scientists who feel this description applies to their own lives appears only to be growing.

## Notes

1 J. Michael Kosterlitz in interview with David Zierler on January 7, 2021, Niels Bohr Library & Archives, American Institute of Physics, College Park, MD USA, https://www.aip.org/history-programs/niels-bohr-library/oral-histories/47183.

2 Giorgio Parisi, "Self-confidence is an important ingredient," Nobel Prize Conversations, Interview February 18, 2022, https://www.nobelprize.org/prizes/physics/2021/parisi/interview/.

3 Pauline Navals, "One on One with Carolyn Bertozzi," *C&EN*, April 8, 2022, vol 100, issue 12, https://cen.acs.org/biological-chemistry/One-on-one-with-Carolyn-Bertozzi/100/i12.

4 Interview of Jim Peebles by Alan Lightman on January 19, 1988, Niels Bohr Library & Archives, American Institute of Physics, College Park, MD USA, https://www.aip.org/history-programs/niels-bohr-library/oral-histories/33957.

5 Thomas Söderqvist, "Introduction: a new look at the genre of scientific biography," in *The History and Poetics of Scientific Biography*, ed. by Thomas Söderqvist (Milton Park: Routledge, 2007), 1–2.

6 According to Söderqvist, 4-5000 biographies of scientists, engineers and medical men and women were published in Latin, French, German, English, Italian, Dutch and the Scandinavian languages in the last 400 years. In the 1990s, circa 75 scientific biographies were published yearly. Söderqvist bases his estimate on Oettinger (1854), Howsam (1997), Morton and More (1994) and *Isis cumulative bibliography*, 1913–65, 1966–1975, 1976–1985 and 1986–1995, in addition, the collections and catalogues of *the Science Museum Library, the Wellcome Library* and *the Royal Society*, and the history of science collection in the *Niedersächsische Staats- und Universitätsbibliothek*, Göttingen. Not included are titles in smaller European or non-European languages. No such estimate of the number of scientific *auto*biographies exists.

7 Steven Shapin, *Scientific Life: A Moral History of a Late Modern Vocation* (Chicago: University of Chicago Press, 2008).

8 Lorraine Daston and H. Otto Sibum, "Introduction: Scientific Personae and Their Histories," *Science in Context* 16 (2003): 1–8.

9 Marlin E. Rice, "Bruce D. Hammock: Science should be fun," *American Entomologist*, 66:1 (2020), 14–19, https://doi.org/10.1093/ae/tmaa010.

10 Ulrich Bröckling, *Postheroische helden. Ein Zeitbild* (Berlin: Suhrkamp, 2020); Ulrich Bröckling, "Heroism and Modernity," https://freidok.uni-freiburg.de/fedora/objects/freidok:151771/datastreams/FILE1/content. See also M. Khan, H. Williams, J. P. Williams and E. French, "Post-Heroic Heroism: Embedded Masculinities in Media Framing of Australian Business Leadership," *Leadership*, 18, no. 2 (2022), 298–327, https://doi.org/10.1177/17427150211049600.

11 Joseph John Thomson, *Recollections and Reflections* (Cambridge: Cambridge University Press, 1936), 6.

12 H. B. G. Casimir, "Physics and Play," in *Haphazard Reality: Half a Century of Science* (New York: Harper & Row, 1983), 125.

13 Victor Weisskopf, *The Joy of Insight: Passions of a Physicist* (New York: Basic Books, 1991), 40.

14 Weisskopf, *The Joy of Insight*, 41.

15 Casimir, *Haphazard Reality*, 125.

16 Weisskopf, *The Joy of Insight*, 49.

17 Andrew Szanton, *The Recollections of Eugene P. Wigner: As Told to Andrew Szanton* (New York: Basic Books, 2003), 105

18 Gerald E. Brown and Sabine Lee, *Hans Albrecht Bethe, Biographic Memoir*, (Washington D.C: National Academy of Sciences, 2009), 8.

19 Walter M. Elsasser, *Memoirs of a Physicist in the Atomic Age* (New York: Science History Publications, 1978), 41.

20 Interview of Werner Heisenberg by Thomas S. Kuhn and John Heilbron on November 30, 1962, Niels Bohr Library & Archives, *American Institute of Physics*, College Park, MD, USA, https://www.aip.org/history-programs/niels-bohr-library/oral-histories/4661-1.

21 Werner Heisenberg, *Physics and Beyond: Encounters and Conversations* (New York: Harper & Row, 1971). See also the memoir of his wife, Elisabeth Heisenberg, *Inner Exile: Recollections of a Life with Werner Heisenberg* (Basel: Birkhauser, 1984). Especially due to his fraught history as a Nazi collaborator.

22 Szanton, *The Recollections of Eugene P. Wigner*, 120, 110.

23 Rebecca Hertzig, *Suffering for Science* (New Brunswick: Rutgers University Press, 2005).

24 Hertzig, *Suffering for Science*, 2–3.

25 Hertzig, *Suffering for Science*, 2–3.

26 Granted, in the nineteenth century a persona appears which puts the individual at the core of science: the scientific genius. One might argue that the scientific genius has older roots – the early eighteenth-century depictions of Isaac Newton come to mind – but the nineteenth century, sometimes called the "heroic century" brought a new, strong attention to the exceptional individual, through three circumstances. First, the focus, in Romantic artistic circles, on the imagination as the fountainhead of original creation. Second, the focus on "great men," through the Scottish historian Thomas Carlyle's 1841 book *On Heroes, Hero-Worship and the Heroic in History,* inspiring many other writers, like Friedrich Nietzsche, Elisabeth Barret Browning, and G. B. Shaw. Third, the interest in inherited high achievement, as charted by Charles Darwin's cousin Francis Dalton in his 1896 book *Hereditary Genius,* a book often said to be the starting point of scientific analysis of genius and later of creativity and high achievement in general.

27 As historians of science have shown, there is a long history of the intermingling between entertainment and science demonstration. Thus, in practice, fun has not been kept as an opposite perspective from transcendence or perseverance as regards science as it could be, because of hundreds of years of science popularizers who have used fun as the "glamour" on top of science to lure non-scientists in (often children but also non-scientist adults) – then one expects fun to turn into joy, which then leads to perseverance and an insight into transcendent values. See Bernhard Lightman, *Victorian Popularizers of Science: Designing Nature for New Audiences* (Chicago: Chicago University Press, 2007).

28 Syukuru Manabe, interview by Adam Smith, nobelprize.org, March 16, 2022. https://www.nobelprize.org/prizes/physics/2021/manabe/185163-manabe-interview-march-2022.

29 Anton Zeilinger, interview by Adam Smith, nobelprize.org, October 4, 2022. https://www.nobelprize.org/prizes/physics/2022/zeilinger/interview/.

30 Neil Kenny, *Curiosity in Early Modern Europe: Word Histories* (Wiesbaden: Harrassowitz, 1998).

31 "Curiosity, n." *OED Online*. April 2023. Oxford University Press. https://www.oed.com/view/Entry/46038?redirectedFrom=curiosity (accessed April 29, 2023).

32 Jon Agar, "2016 Wilkins–Bernal–Medawar lecture: The Curious History of Curiosity-driven Research" *Notes and Records* 71, (2017), 409–429, http://doi.org/10.1098/rsnr.2017.0034.

33 Sophia Vasalou, *Wonder: A Grammar* (New York: State University of New York Press, 2016); Lorraine Daston and Katharine Park, *Wonders and the Order of Nature: 1150–1750* (Cambridge, MA: MIT Press/Zone Books, 1998).

34 *Life Magazine,* May 2, 1955, personal memoir of William Miller, through Alice Calaprice, *The Quotable Einstein, with a Foreword by Freeman Dyson* (New Jersey: Princeton University Press, 1996).

35 For an example of the American popular press describing Einstein as a sage, see "Can We Produce an Einstein?" Editorial *Life*, May 2, 1955, 42. Aaron Lecklider, *Inventing the Egghead: The Battle over Brainpower in American Culture* (University of Pennsylvania Press, Philadelphia, 2013) also has a chapter on the depiction of Einstein in the United States.

36 Torrey Levin-Russell, "Richard Feynman," https://myhero.com/Richard_Feynman_US.

37 Albert Einstein, Letter to Carl Seelig, March 11, 1953, Einstein Archive 39-013, through Alice Calaprice, *The Quotable Einstein* (Princeton: Princeton University Press, 2013), 12.

38 Szanton, *The Recollections of Eugene P. Wigner*, 122.

39 Szanton, *The Recollections of Eugene P. Wigner*, 24.

40 Casimir, *Haphazard Reality.*

41 Kimmel, *Manhood in America.*

42 Margaret Connolly, *The Life Story of Orison Swett Marden – A Man Who Benefited Men Marden* (New York City: Thomas Y. Crowell Company), 1925.

43 Timothy Hampton, *Cheerfulness: A Literary and Cultural History* (Princeton: Zone Books/Princeton University Press, 2022); Christina Kotchemidova, "From Good Cheer to 'Drive-by-Smiling': A Social History of Cheerfulness," *Journal of Social History*, 39, no. 1, (2005), 5–37.

44 This was also eagerly encouraged by growing American consumer culture promoting the visiting of amusement parks like Coney Island. See Roger Horowitz, ed., *Boys and Their Toys: Masculinity, Class and Technology in America* (New York, Routledge, 2001).

45 Adam Potkay, *The Story of Joy: From the Bible to Late Romanticism* (Cambridge: Cambridge University Press, 2007).

46 Google Ngram Viewer, search word "fun," https://books.google.com/ngrams/graph?content=fun&year_start=1800&year_end=2019&corpus=en-2019&smoothing=0.

47 Linus Torvalds, *Just for Fun: The Story of an Accidental Revolutionary* (New York: Harper Business, 2002).

48 Ulrich Bröckling, "Negations of the Heroic: A Typological Essay" in Nicole Falkenhayner, Sebastian Meurer, Tobias Schlectriemen, "Analyzing Processes of Heroization: Theories, Methods, Histories," *Helden. Heroes. Héros. E-journal zu Kulturen des Heroischen*, 5, 2019, (10.6094/helden.heroes.heros./2019/APH/12); Ulrich Bröckling, *Heroism and Modernity*, 1.

49 Edward Teller, "Should Your Child Be a Scientist?" advertisement for New York Life Insurance company, *Scientific American* 198, no 5 (1958), 141–142.

50 Abraham Maslow, *The Farther Reaches of Human Nature* (London: Penguin, 1993 [1971]), 59.

51 Hampton, *Cheerfulnes*, 10–23.

52 Richard Feynman, *Fun to Imagine,* BBC broadcast July 8–August 12, 1983, https://www.bbc.co.uk/programmes/p0198zc1.

53 On January 28, 1986, on live television, the space shuttle Challenger exploded, killing its entire crew. Describing his experiences on the NASA committee, Feynman says he was initially reluctant to join because "I won't be able to work with this physics problem I've been having fun with." Richard P. Feynman, "An Outsider's Insider View of the Challenger Inquiry," *Physics Today*, (1988), 26.

54 Richard P. Feynman & Ralph Leighton (ed.), *Surely You're Joking, Mr. Feynman! Adventures of a Curious Character* (New York: W. W. Norton & Co,1985); Richard P. Feynman & Ralph Leighton (ed.), *What Do You Care What Other People Think? Further Adventures of a Curious Character* (New York: W. W. Norton & Co, 1988).

55 *Surely You're Joking, Mister Feynman: Adventures of a Curious Character* is number 20 on the best-seller list over scientific memoirs on American Amazon in April 2023, 39 years after publication.

56 Edwin M. Yoder Jr., "A physicist's guide to the galaxy," *The Washington Post*, February 17, 1985. https://www.washingtonpost.com/archive/entertainment/books/1985/02/17/a-physicists-guide-to-the-galaxy/0e23df46-02d4-4bb9-8227-02e7ce7bc5e2/.

57 K. C. Cole, "Pranks of a Nobel laureate, *New York Times*, Late City Final Edition, January 27, 1985, https://archive.nytimes.com/www.nytimes.com/books/97/09/21/reviews/feynman-joking.html.

58 Feynman, *Surely you're joking, mister Feynman!*, 173. Italics in original.

59 Feynman, *Surely you're joking, mister Feynman!*, 174.

60 Feynman, *Surely you're joking, mister Feynman!*, 174.

61 Leila McNeill, "Surely you're a creep, Mr. Feynman," *The Baffler* 43, (2019), https://thebaffler.com/outbursts/surely-youre-a-creep-mr-feynman-mcneill.

62 Anja Skaar Jacobsen, review of Jörg Resag, *Feynman and his Physics: The Life and Science of an Extraordinary Man* (Springer, 2018), *Isis*, 111, no. 4, (2020), 899–900.

63 Torrey Levin-Russell, "Richard Feynman," https://myhero.com/Richard_Feynman_US.

64 Torrey Levin-Russell, "Richard Feynman," https://myhero.com/Richard_Feynman_US.

65 James Gleick, "Richard Feynman Dead at 69; Leading Theoretical Physicist," *The New York Times*, February 17, 1988. https://www.nytimes.com/1988/02/17/obituaries/richard-feynman-dead-at-69-leading-theoretical-physicist.html.

66 The only critical voice directed towards Feynman from within the physics community comes from Murray Gell-Mann, who claimed that Feynman was too focused on self-presentation, and from John Clauser, Nobel laureate in physics 2022: "I didn't particularly care for Feynman. I didn't really like him very much. And frankly a lot of the stuff that he did and said was wrong, especially in foundations of quantum mechanics." Interview of John Clauser by Joan Bromberg on May 20, 2002, Niels Bohr Library & Archives, *American Institute of Physics*, College Park, MD USA, www.aip.org/history-programs/niels-bohr-library/oral-histories/25096.

67 Christopher Sykes, *No Ordinary Genius: The Illustrated Richard Feynman* (New York: W. W. Norton, 1994), 54.

68 James Gleick, *Genius: The Life and Science of Richard Feynman* (New York: Pantheon Books, 2011[1992]), 8.

69 See for instance George Johnson, "The Jaguar and the Fox," *The Atlantic*, July 2000, https://www.theatlantic.com/magazine/archive/2000/07/the-jaguar-and-the-fox/378264/.

70 Erika Lorraine Milam and Robert A. Nye, "An Introduction to Scientific Masculinities," *Osiris*, 30, (2015) 1–14; Michael Kimmel, *Manhood in America: A Cultural History* (Oxford University Press, Oxford, 2012).

71 Lecklider, *Inventing the Egghead.*

72 Weisskopf, *Joy of Insight*, 37–38.

73 Szanton, *The Recollections of Eugene P. Wigner*, 125.

74 Potkay, *The Story of Joy*, 73–95.

75 Jennifer Clement, "Passions and Preaching: The Early Modern English Sermon, 1603–1660," unpublished manuscript.

76 Monique Scheer, *Enthusiasm* (Oxford: Oxford University Press, 2020).

77 Oxford English Dictionary, search word "fun,“ the first definition being "An act of fraud or deception; a trick played on a person; a joke," going back to 1699 B. E. *New Dict. Canting Crew* "*Fun*, a Cheat or slippery Trick." See also Merriam Webster Dictionary, search word "fun," which defines fun as "to hoax, perhaps alteration of Middle English *fonnen*, from *fonne* dupe," first mentioned 1726, https://www.merriam-webster.com/dictionary/fun#word-history:

78 Mihaly Csikszentmihalyi, *Beyond Boredom and Anxiety: Experiencing Flow in Work and Play* (San Francisco: Jossey-Bass, 1975).

79 Shapin, *Scientific Life.*

80 Michel Foucault, "Technologies of the Self," Lectures at University of Vermont, October 1982, in *Technologies of the Self* (Amherst: University of Massachusetts Press, 1988), 16–49.

81 Hertzig, *Suffering for Science*, 2–3.

82 Samuel Franklin, *The Cult of Creativity* (Chicago: University of Chicago, 2023); Bregje van Eekelen, "Creative Intelligence and the Cold War: US Military Investments in the Concept of Creativity, 1945–1965," *Conflict and Society*, 3, no. 1 (2017): 92; Jamie Cohen-Cole, *The Open Mind: Cold War Politics and the Sciences of Human Nature* (Chicago: Chicago University Press, 2014).

83 Donna Strickland, Nobel Banquet Speech, December 10, 2018 https://www.nobelprize.org/prizes/physics/2018/strickland/speech/.

84 Interview of John Clauser by Joan Bromberg on May 20, 2002, Niels Bohr Library & Archives, American Institute of Physics, College Park, MD USA, https://www.aip.org/history-programs/niels-bohr-library/oral-histories/25096.

85 Robert Iliffe, *Priest of Nature: The Religious Worlds of Isaac Newton* (Oxford: Oxford University Press, 2017).

86 According to *Merriam-Webster*, the first use of the word "egghead" was in 1907, and it described *scientists* specifically. According to Google Books Ngram Viewer "egghead" was a hugely popular word from 1948–1959, becoming more than nine times more prevalent in only ten years, then went out of fashion almost as quickly as it rose. Also, see Lecklider, *Inventing the Egghead.*

## Bibliography


Agar, Jon. "2016 Wilkins–Bernal–Medawar lecture: The Curious History of Curiosity-driven Research" *Notes and Records* 71 (2017): 409–429, https://doi.org/10.1098/rsnr.2017.0034.

Bröckling, Ulrich. "Negations of the Heroic: A Typological Essay." In *Analyzing Processes of Heroization. Theories, Methods, Histories*, edited by Nicole Falkenhayner, Sebastian Meurer and Tobias Schlechtriemer. *Helden. Heroes. Héros. E-journal zu Kulturen des Heroischen* (2019): https.doi.org/10.6094/helden.heroes.heros./2019/APH/12.

Bröckling, Ulrich. "Heroism and Modernity." Albert-Ludwigs-Universität Freiburg, 2019. https://freidok.uni-freiburg.de/fedora/objects/freidok:151771/datastreams/FILE1/content

Ulrich Bröckling. *Postheroische helden. Ein Zeitbild.* Berlin: Suhrkamp, 2020.

Brown, Gerald E. and Sabine Lee. *Hans Albrecht Bethe, Biographic Memoir.* Washington D.C: National Academy of Sciences, 2009.

Casimir, H. B. G. *Haphazard Reality: Half a Century of Science.* New York: Harper & Row, 1983.

Clauser, John interview by Joan Bromberg on May 20, 2002. Niels Bohr Library & Archives, American Institute of Physics, College Park, MD USA, https://www.aip.org/history-programs/niels-bohr-library/oral-histories/25096.

Clement, Jennifer. "Passions and Preaching: The Early Modern English Sermon, 1603–1660," unpublished manuscript.

Cohen-Cole, Jamie. *The Open Mind: Cold War Politics and the Sciences of Human Nature.* Chicago: Chicago University Press, 2014.

Cole, K. C. "Pranks of a Nobel laureate," *New York Times*, Late City Final Edition, January 27, 1985, https://archive.nytimes.com/www.nytimes.com/books/97/09/21/reviews/feynman-joking.html

Connolly, Margaret. *The Life Story of Orison Swett Marden – A Man Who Benefited Men.* New York City: Thomas Y. Crowell Company, 1925.

Csikszentmihalyi, Mihaly. *Beyond Boredom and Anxiety: Experiencing Flow in Work and Play.* San Francisco: Jossey-Bass, 1975.

Daston, Lorraine and Katharine Park. *Wonders and the Order of Nature: 1150–1750.* Cambridge, MA: MIT Press/Zone Books, 1998.

Daston, Lorraine and H. Otto Sibum. "Introduction: Scientific Personae and Their Histories," *Science in Context* 16 (2003): 1–8.

Edwin M. and Yoder Jr. "A physicist's guide to the galaxy." *The Washington Post*, February 17, 1985. https://www.washingtonpost.com/archive/entertainment/books/1985/02/17/a-physicists-guide-to-the-galaxy/0e23 df46-02d4-4bb9-8227-02e7ce7bc5e2/.

Einstein, Albert. Letter to Carl Seelig, March 11, 1953, Einstein Archive 39–013, through Alice Calaprice, *The Quotable Einstein*, 12. Princeton: Princeton University Press, 2013.

Elsasser, Walter M. *Memoirs of a Physicist in the Atomic Age.* New York: Science History Publications, 1978.

Feynman, Richard P. "An Outsider's Insider View of the Challenger Inquiry," *Physics Today*, (1988): 26–37.

Feynman, Richard P. and Ralph Leighton (ed.). *Surely You're Joking, Mr. Feynman! Adventures of a Curious Character.* New York: W. W. Norton & Co, 1985.

Feynman, Richard P. and Ralph Leighton (ed.). *What Do You Care What Other People Think? Further Adventures of a Curious Character.* New York: W. W. Norton & Co, 1988.

Feynman, Richard. *Fun to Imagine,* BBC broadcast July 8–August 12, 1983, https://www.bbc.co.uk/programmes/p0198zc1.

Foucault, Michel. "Technologies of the Self." Lectures at University of Vermont, October 1982. In *Technologies of the Self.* Amherst: University of Massachusetts Press, 1988, 16–49.

Franklin, Samuel. *The Cult of Creativity.* Chicago: University of Chicago, 2023.

Gleick, James. "Richard Feynman Dead at 69; Leading Theoretical Physicist." *The New York Times*, February 17, 1988. https://www.nytimes.com/1988/02/17/obituaries/richard-feynman-dead-at-69-leading-theoretical-physicist.html.

Gleick, James. *Genius: The Life and Science of Richard Feynman.* New York: Pantheon Books, 2011[1992].

Hampton, Timothy. *Cheerfulness: A Literary and Cultural History.* Princeton: Zone Books/Princeton University Press, 2022.

Heisenberg, Elisabeth. *Inner Exile: Recollections of a Life with Werner Heisenberg.* Basel: Birkhäuser, 1984.

Heisenberg, Werner interview by Thomas S. Kuhn and John Heilbron on November 30, 1962. Niels Bohr Library & Archives, *American Institute of Physics*, College Park, MD USA, https://www.aip.org/history-programs/niels-bohr-library/oral-histories/4661-1.

Heisenberg, Werner. *Physics and Beyond: Encounters and Conversations.* New York: Harper & Row, 1971.

Hertzig, Rebecca. *Suffering for Science.* New Brunswick: Rutgers University Press, 2005.

Horowitz, Roger, ed. *Boys and Their Toys: Masculinity, Class and Technology in America.* New York, Routledge, 2001.

Iliffe, Robert. *Priest of Nature: The Religious Worlds of Isaac Newton.* Oxford: Oxford University Press, 2017.

Johnson, George. "The Jaguar and the Fox," *The Atlantic* (July 2000), https://www.theatlantic.com/magazine/archive/2000/07/the-jaguar-and-the-fox/378264/.

Kenny, Neil. *Curiosity in Early Modern Europe: Word Histories.* Wiesbaden: Harrassowitz, 1998.

Khan, M., H. Williams, J. P. Williams and E. French. "Post-Heroic Heroism: Embedded Masculinities in Media Framing of Australian Business Leadership," *Leadership* 18, no 2 (2022): 298–327. https://doi.org/10.1177/17427150211049600.

Kimmel, Michael. *Manhood in America: A Cultural History.* Oxford: Oxford University Press, 2012.

Kosterlitz, J. Michael interview by David Zierler on January 7, 2021. Niels Bohr Library & Archives, American Institute of Physics, College Park, MD USA, https://www.aip.org/history-programs/niels-bohr-library/oral-histories/47183.

Kotchemidova, Christina. "From Good Cheer to 'Drive-by-Smiling': A Social History of Cheerfulness," *Journal of Social History* (2005): 5–37.

Lecklider, Aaron. *Inventing the Egghead: The Battle over Brainpower in American Culture.* Philadelphia: University of Pennsylvania Press, 2013.

Levin-Russell, Torrey. "Richard Feynman," https://myhero.com/Richard_Feynman_US.

Lightman, Bernhard. *Victorian Popularizers of Science: Designing Nature for New Audiences.* Chicago: Chicago University Press, 2007.

Manabe, Syukuru interview by Adam Smith, nobelprize.org, March 16, 2022. https://www.nobelprize.org/prizes/physics/2021/manabe/185163-manabe-interview-march-2022.

Maslow, Abraham. *The Farther Reaches of Human Nature.* London: Penguin, 1993 [1971].

McNeill, Leila. "Surely you're a creep, Mr. Feynman," *The Baffler* 43 (2019), https://thebaffler.com/outbursts/surely-youre-a-creep-mr-feynman-mcneill.

Milam, Erika Lorraine and Robert A. Nye. "An Introduction to Scientific Masculinities," *Osiris* (2015): 1–14.

Miller, William. "Personal memoir." *Life Magazine,* May 2, 1955. In Alice Calaprice, *The Quotable Einstein, With a Foreword by Freeman Dyson.* New Jersey: Princeton University Press, 1996.

Navals, Pauline. "One on One with Crolyn Bertozzi," *C&EN,* April 8, 2022, https://cen.acs.org/biological-chemistry/One-on-one-with-Carolyn-Bertozzi/100/i12.

*Oxford English Dictionary Online.* Oxford: Oxford University Press, April 2023.

Parisi, Giorgio. "Self-confidence is an important ingredient," Nobel Prize Conversations, Interview February 18, 2022, https://www.nobelprize.org/prizes/physics/2021/parisi/interview/.

Peebles, Jim interview by Alan Lightman on January 19, 1988. Niels Bohr Library & Archives, American Institute of Physics, College Park, MD USA, https://www.aip.org/history-programs/niels-bohr-library/oral-histories/33957.

Potkay, Adam. *The Story of Joy: From the Bible to Late Romanticism.* Cambridge: Cambridge University Press, 2007.

Resag, Jörg. "Feynman and his Physics: The Life and Science of an Extraordinary Man." Review by Anja Skaar Jacobsen in *Isis* 2018: 899–900.

Rice, Marlin E. "Bruce D. Hammock: Science should be fun," *American Entomologist* 66(1) (2020): 14–19. https://doi.org/10.1093/ae/tmaa010.

Scheer, Monique. *Enthusiasm.* Oxford: Oxford University Press, 2020.

Shapin, Steven. *Scientific Life: A Moral History of a Late Modern Vocation.* Chicago: University of Chicago Press, 2008.

Söderqvist, Thomas. "Introduction: a new look at the genre of scientific biography." In *The History and Poetics of Scientific Biography,* edited by Thomas Söderqvist, 1–2. Milton Park: Routledge, 2007.

Strickland, Donna. Nobel Banquet Speech, December 10, 2018 https://www.nobelprize.org/prizes/physics/2018/strickland/speech/.

Sykes, Christopher. *No Ordinary Genius: The Illustrated Richard Feynman.* New York: W. W. Norton, 1994.

Szanton, Andrew. *The Recollections of Eugene P. Wigner: As Told to Andrew Szanton.* New York: Basic Books, 2003.

Teller, Edward. "Should Your Child Be a Scientist?" advertisement for New York Life Insurance company, *Scientific American* 198, no. 5 (1958), 141–142.

Thomson, Joseph John. *Recollections and Reflections.* Cambridge: Cambridge University Press, 1936.

Torvalds, Linus. *Just for Fun: The Story of an Accidental Revolutionary.* New York: Harper Business, 2002.

Ulrich Bröckling. *Heroism and Modernity*, https://freidok.uni-freiburg.de/fedora/objects/freidok:151771/datastreams/FILE1/content.

van Eekelen, Bregje. "Creative Intelligence and the Cold War: US Military Investments in the Concept of Creativity, 1945–1965," *Conflict and Society* 3, no. 1 (2017): 92.

Vasalou, Sophia. *Wonder: A Grammar.* New York: State University of New York Press, 2016.

Weisskopf, Victor. *The Joy of Insight: Passions of a Physicist.* New York: Basic Books, 1991.

Edwin M. Yoder Jr. "A physicist's guide to the galaxy," *The Washington Post*, February 17, 1985. https://www.washingtonpost.com/archive/entertainment/books/1985/02/17/a-physicists-guide-to-the-galaxy/0e23df46-02d4-4bb9-8227-02e7ce7bc5e2/.

Zeilinger, Anton interview by Adam Smith, nobelprize.org, October 4, 2022. https://www.nobelprize.org/prizes/physics/2022/zeilinger/interview/

CHAPTER SIX

# Demythologizing Science: Reijer Hooykaas on Hero Worship as "Undesirable" and "Disdaining"

*Jelmer Heeren*

> Neither blame nor praise, but understanding is needed.
>
> —Reijer Hooykaas[1]

**Abstract**

For historians of science, the question of how to write about 'heroes' in science is related to other methodological questions: What is science? What is 'good' science and therefore a 'good' scientist? What phenomena are worthy of study? What is the relationship between an individual and his or her context? The issue of 'heroes' specifically occasioned a watershed between many early twentieth-century historians of science and earlier generations of (lay) historians. Reijer Hooykaas (1906-1994) was among this new generation of historians. I describe and explain the changing ways in which, broadly speaking, twentieth-century historians of science have conceived of 'heroes' in science. I use Hooykaas as a looking glass through which to describe the broader historiographical developments.

**Keywords:** Historiography of science, Reijer Hooykaas, hero worship, genius, 20th century

Hendrik Bakhuis Roozeboom (1854-1907) passed away at the age of 52, before finishing a multi-volume book series on the phase behavior of substances. Praised by his fellow chemists worldwide, the Dutchman's work allowed for advancements in chemistry, geology, and metallurgy – including research on the different types of steel – without which the technical developments of the first half of the twentieth century "would not have taken such a high flight."[2] As the news broke of his passing, chemist Andreas Smits (1870-1948) speculated that if he had lived longer, Bakhuis Roozeboom "would undoubtedly have received high esteem from all sides." Years later, former-student Hugo Kruyt (1882-1959) likewise lamented that Bakhuis Roozeboom's inability to finish what he started "had disastrous consequences" for the "recognition in the world of its importance." Physicist J. Fokkens (birth and death years unknown) speculated he perhaps would have received a Nobel Prize, were it not for his untimely death.[3] In contrast, Bakhuis Roozeboom's contemporaries Henri van 't Hof (1852-1911), Hendrik Lorentz (1853-1928) and Heike Kamerlingh Onnes (1853-1926) became Nobel laureates.

In 1953, the *Rijksmuseum voor de Geschiedenis der Natuurwetenschappen* (today called Rijksmuseum Boerhaave) in Leiden commemorated the centennials of the birthdays of Kamerlingh Onnes and Lorentz. A comprehensive exhibition succeeded in drawing larger crowds than usual. Cementing the image of these scientists as geniuses, newspaper coverage of the exhibition made note of "the remarkable giftedness, traits and feats of both physicists, achieved through hard labour, vision and perseverance (Onnes) or, even more stunningly, without any apparent effort (Lorentz)."[4] By comparison, a year later, the commemoration activities surrounding the centenary of Bakhuis Roozeboom's birthday were quite modest. Three professional associations – including *the Koninklijke Nederlandse Akademie van Wetenschappen* (KNAW, Royal Netherlands Academy of Arts and Sciences) – organized several activities, including a three-day exhibition in one of the laboratory rooms of the Universiteit van Amsterdam. These activities targeted a specialist and therefore niche audience and did not draw large crowds.[5]

However, there was another place where Bakhuis Roozeboom was commemorated. The day before the centennial of Bakhuis Roozeboom's birthday, science historian Reijer Hooykaas (1906-1994) gave a speech during a regular meeting of the *Christelijke Vereniging van Natuur- en Geneeskundigen in Nederland* (Christian Society of Scientists and Physicians in the Netherlands) in one of the classrooms of the physics laboratory of the then Calvinist Vrije Universiteit Amsterdam.[6] Characteristic of Hooykaas, he did not belabour the unique giftedness and specialist achievements of the chemist. Instead, Hooykaas showed how Bakhuis Roozeboom's science and his Christian faith were related. He described how this "humble man" (*bescheiden man*) experienced "great joy" (*grote vreugde*) in doing research and teaching and found in Bakhuis Roozeboom an example of how science in general, chemistry in particular, reveals both humanity's "lofty greatness" (*verheven grootheid*) and its "insignificant smallness" (*onbeduidende kleinheid*).[7] In sum, rather than praising *remarkable giftedness* – as was the case with the popular coverage of the Kamerlingh Onnes/Lorentz commemoration the year before – Hooykaas praised Bakhuis Roozeboom's *attitude of humility*.

This example illustrates that scientists may be commemorated in different ways and for different reasons. It also highlights that "popular" and "professional" remembrance may diverge from one another. For historians of science, the question of how to relate to "heroes" in science is part of a set of questions – for some today unanswerable at best and unhelpful at worst – that (ultimately) informs their historiography: what is science? What is "good" science and therefore a "good" scientist? What historical phenomena and actors are worthy of studying? And to what degree is the individual scientist a product of his or her context? Interrelated as these questions are, the issue of "heroes" specifically occasioned a watershed between many early twentieth-century historians of science and earlier generations of (lay) historians. Hooykaas was among this new generation of historians and the first in the Netherlands to hold a chair in the history of science, with professorships at the Vrije Universiteit Amsterdam (1945-1971) and the Rijksuniversiteit Utrecht (1967-1976).[8]

In this paper, I describe and explain the changing ways in which, broadly speaking, twentieth-century historians of science have conceived of "heroes" in science. I take the term "heroes" loosely to mean anything between people that deserve history writing, role models, people to look up to, worthy of emulation and people that are portrayed to have contributed to science to such a degree that they seem almost otherworldly. There are stronger and weaker definitions at play here and both ends of the spectrum will be relevant in this paper. To both limit the scope of this article and to provide specific examples, I will proceed by using Hooykaas as a looking glass by which I describe the broader developments in the historiography of science. I will (1) describe the way in which Hooykaas distanced himself from the historiography of genius in his early years as a historian of science. I will (2) illustrate this by explicating the way Hooykaas treated eighteenth-century chemist Antoine Lavoisier, a French hero in the history of science. Next (3) it is necessary to give an overview of Hooykaas' philosophy of science to demonstrate how philosophical and historiographical presuppositions result in different perspectives on who are considered "heroes" (and why) and who are not. Consequently, the changing philosophical and historiographical landscape in the second half of the twentieth century allows me to conclude with sections on the way further contextualization efforts raised questions about (4) the supposed special nature of science and (5) the role of individuals in science.

## Against the historiography of genius

Trained as a chemist in Utrecht in the latter half of the 1920s, Hooykaas wanted to write a historical work to obtain a doctoral degree in his field. Doing so in such an experimental research-oriented environment as the Faculty of Mathematics and Natural Science at the Rijksuniversiteit Utrecht was highly unusual. The institutionalization and professionalization of the history of science as a discipline had barely started worldwide. Hooykaas' chemistry teacher Ernst J. Cohen (1889-1944) was historically minded but only in a so-called positivist sense – on the walls of his laboratory, named after Nobel Prize laureate Henri van 't Hoff, were hung portraits of famous old chemists. Although Cohen himself did original historical research by consulting primary sources, he did not care for the historical context of science.[9]

Cohen was not unique in his approach to history: virtually all nineteenth and early twentieth century historians of chemistry deemed chemistry to be the "positive" science *par excellence*. They maintained that science provides certain knowledge over and against religious, philosophical or any other metaphysical speculation. If anything, obstacles to science always come from the outside: externalities that interfere with the unique scientific rationality. Predicated on the notion of progress, there developed a positivist-Whig historiography, as it would later be called, in which past thinkers were divided into two groups: the correct, scientifically open-minded observers over and against the incorrect, superstitious and dogmatic speculators. On this account, the heroes are those that made lasting contributions to the body of knowledge that the present shows to be true. That is, the heroes in science are geniuses, gifted individuals who – although hindered by a number of external factors – discover, often through eureka-moments, an existing unity between thought and world. Typically, Whig-positivist historians did not pay attention to the historical circumstances and focused only on the scientist's contributions, placed within a series of other discoveries, revelatory moments and achievements by other heroes that all led to true scientific knowledge.[10]

In the case of chemistry, these early historians and chemists declared the French chemist Antoine Lavoisier (1743-1794) to be their hero: he "did not [merely] revolutionize chemistry, he founded it."[11] Their histories were accounts of the triumph of Lavoisier over his pre-scientific, irrational colleagues: the wellspring from which true chemistry flowed. Whether it was through his experiments or clear-headed thinking, Lavoisier was chemistry's hero.[12] Not everyone saw it this way, however: some English writers favored Robert Boyle (1627-1691) and some Germans let chemistry start with George Ernst Stahl (1659-1734). Historiography of science in the Netherlands during this time too was infused with nationalist sentiments. Herman Boerhaave (1668-1738) and Christiaan Huygens (1629-1695) were the Dutch heroes of science and efforts were made to raise statues in their honour. The early meetings of the *Vereeniging voor Geschiedenis der Genees-, Natuur- en Wiskunde* (today called Gewina) were often organized around memorials of heroes in the history of medicine including Clusius (1526-1609) and Vesalius (1514-1564). A commemoration of Boerhaave in 1918 included the attendance of the minister of Education and other notables.[13]

Although nineteenth and early twentieth-century historiography of science centered around heroes, this does not mean that no interesting histories were written. "Every commemoration encouraged the panegyrists to conduct original historical research," historian Klaas van Berkel observes.[14] Yet, in the Netherlands, it was only with the writings of Eduard Jan Dijksterhuis (1892-1965) that a radically different approach to the history of science was put forward. In his *Val en worp* (1924) he took seriously Pierre Duhem's thesis that there was much more continuity between the Middle Ages and the Scientific Revolution of the seventeenth century than the Whig-positivists assumed. More than Duhem, however, he emphasized the importance of studying the past in its context and avoiding judging it based on today's standards.[15] It was along these lines – against the historiography of genius – that Hooykaas wanted to pursue the history of his field and write his dissertation entitled *Het begrip element in zijn historisch-wijsgeerige ontwikkeling* (The Concept of Element in Its Historical-Philosophical Development).[16]

Hooykaas found in chemist Hugo Kruyt – as mentioned above himself a student of Bakhuis Roozeboom and of Cohen – a willing doctoral supervisor, although he had no experience with historical research.[17] Hooykaas taught himself how to do history of science and drafted a basic framework of his method in the introduction to his dissertation, which covers the evolution and philosophical contexts of the concept of element from Aristotle to Hooykaas' own time. In this introduction he insists that we must avoid judging the past in light of the present; to understand the past is to approach it on its own terms. "It is foolishness to think," he submits, "that the old chemists were somewhat backward in their thinking; they approached their problems as logically or illogically as we would." A sense of humility is appropriate as history shows that scientific truths are relative to their time: "We will have to acknowledge that ideas that now seem foolish, in their time expressed the truth [...] we do not deny the advancement of science, but the path leading to the Tree of Knowledge is winding and often turns on itself [...] this makes us less triumphant and humbler with respect to our ancestors."[18] Historian of science (and student of Hooykaas) Floris Cohen would later write that Hooykaas' introduction to his dissertation can be viewed alongside other works from the same period articulating "rules – or rather [...] a mindset – required for turning the history of science into real history."[19]

## The case of Antoine Lavoisier

As mentioned above, the figure of Lavoisier loomed large in the history of chemistry. Accordingly, Hooykaas immediately confronts the reader of his dissertation with the implications of the new historiography for the place of the French chemist:

> With a self-conscious certainty that our chemistry was the truth and Lavoisier its prophet, we could only condemn the Ancients. Because science had simply 'proved' that the elements could not be converted into each other, showing the foolishness of the alchemists'

> attempts. In this way the history of chemistry became the history of an error, in which every now and then light can be found whenever a correct insight is had. It is those lights that constitute the actual history of our science. As far as one deals with alchemist theories, it is done as a pastime after laborious laboratory work: one writes a 'histoire anecdotique' and allows the light of the nineteenth century to shine brighter against the dark background of the Middle Ages.[20]

Hooykaas' shift away from Lavoisier as all-important resonates throughout the work. All in all, he argues that there is a high degree of continuity between Lavoisier and his predecessors – alchemists among them – in contrast with the Whig-positivist historiography that had declared Lavoisier the genius founder of chemistry, a field of study free from the shackles of religious and philosophical speculation.

Hooykaas was not the first to point out the continuity between Lavoisier and his predecessors. In his dissertation, he draws on the French science historians Pierre Duhem (1861-1916), Émile Meyerson (1859-1933), and Hélène Metzger (1889-1944) who had been engaged in "hot debates" about Lavoisier. These debates were the occasion for them to "experiment with a new anti-positivist attitude through a reappraisal of outdated and rejected theories."[21] Accordingly, Metzger's work had already evoked reaction a decade before Hooykaas published his dissertation. The prominent positivist historian of science George Sarton (1884-1956), considered by many to be the main actor behind the professionalization and institutionalization of the discipline of the history of science, had criticized Metzger's 1923 *Les doctrines chimiques* for divorcing the development of chemical doctrines from the life of scholars. In her desire to avoid the writing of a history of great men, Metzger had focused on chemistry's doctrines, not on the chemists. She used their names only for periodization purposes.[22] In contrast, a consideration of the scholars' biographies would, Sarton countered, lead to a richer and more human understanding of the science. He asked:

> To what social, religious, intellectual, geographical, professional milieu did they belong? All this is very important to know and we cannot weigh their words if we do not know anything about them. Isn't her [i.e. Metzger's] goal to make us acquainted with the intellectual atmosphere of scientific doctrines, but how can she succeed in this without introducing us to the men who lived there and who constantly transformed it?[23]

A similar critique could have been leveled against Hooykaas' dissertation: he, too, favors the ideas in their historical and philosophical context above the men's lives.[24] For Sarton, true to his positivist outlook, a more "human" understanding precisely entailed the emphasis on those men who – despite the erroneous ways of their age – had prevailed and advanced science: "The mediocrity of the milieu brings out the greatness of the heroes."[25] In contrast, Hooykaas follows in Metzger's footsteps not only by making the heroes' achievements and ideas relative, but also by rehabilitating the alchemists.

In the following years, Hooykaas showed he was not beholden to a historiography that focused solely on the content of science as it developed through time. He found a way to satisfy the Sartonian desire for biography without invoking Whig-positivism. In fact, already in 1930 he became interested in Blaise Pascal (1623-1662) and Boyle, which led to publications in 1939 and 1942 respectively – the latter of which Michael Hunter and John Hedley Brooke called a "magnum opus."[26] In his work on Boyle, Hooykaas intentionally chooses not to discuss either Boyle's experiments and explanations or his attempts to correct his predecessors. Where the older historiography would try to identify Boyle's contributions to today's "true" and "positive" science, Hooykaas situates Boyle's theories and ideas in their philosophical context, providing comparisons with contemporaries and showing how his theology and science are inseparably linked in terms of their common conception of experience and reason.[27]

So, when Hooykaas returned to Lavoisier with a 60-page essay in 1952,[28] he did so by tracing Lavoisier's unity of thought, as manifested throughout his writings and activities, in connection with broader intellectual developments in the eighteenth century. The content of Lavoisier's chemistry and its relationship to that of his contemporaries and predecessors is discussed in under 20 pages; the other 40 pages are devoted to intellectual trends, Lavoisier's involvement with improving the tax collecting system, agriculture and educational policy and the manufacturing of gunpowder, his finances, his wife's involvement with his science, his involvement with the French Academy of Sciences, and the circumstances that led up to his arrest and execution in 1794 under the oversight of the revolutionary regime, among other things.[29] Hooykaas' evaluation of Lavoisier's place in the history of chemistry thereby stays fundamentally the same: expanding and adding to earlier expressed sentiments, Hooykaas again demonstrates Lavoisier to be truly beholden to his predecessors, privy to their "judgments and prejudices."[30] For example, Hooykaas points out that Lavoisier's new theories were largely dependent on the "groundbreaking" (*baanbrekend*) experimental results provided to him by the likes of Joseph Priestley (1733-1804) and Henry Cavendish (1731-1810).[31]

True to his aim, Hooykaas traces the way Lavoisier's thought relates to the latter half of the eighteenth century, an age of optimism and of revolution, an age of leaving the past behind through an unwavering confidence in Reason, education and social programs. His confidence in "Reason," Hooykaas argues, seeps into Lavoisier's chemistry, despite his adherence to the empirical, anti-Cartesian, ideas of such thinkers as Condillac (1714-1780). Lavoisier is not true to his own empirical confession, Hooykaas points out. The law of the conservation of energy was for instance not experimentally demonstrated by Lavoisier: he assumed it *a priori*. Hooykaas portrays Lavoisier as both beholden to the spirit of his age and yet possessing an independent mind. Hooykaas praises Lavoisier's attempt to rescue the French Academy of the Sciences during the French Revolution from the purge of anyone remotely royalist; he seemingly never lets go of his ideal that humanity must be educated in the new natural science. His faith in "Progress" and the "Light" now acquired, through "Reason," remains unshaken until the end, Hooykaas submits. Only when the Revolution enters a more violent stage and the passions take hold of the populace, only when Lavoisier himself is wrongly imprisoned, does he realize that a "disease" has destroyed France. It is with this tension Hooykaas concludes his essay but not before he himself weighs in, offering his own take on this tension:

> A vacuum must have arisen for Lavoisier; his brilliant career has ended in the disappointment in his belief in man's innate or reason-acquired goodness. Had he based his view of life, not on philosophy, but on that religion, "which is true because it has well known our nature" (Pascal), he might perhaps have been less disappointed.[32]

### Hooykaas' hero: Blaise Pascal

Ending the essay with such a personal take – all the while invoking Blaise Pascal who lived a century before Lavoisier – strikes the contemporary historian as odd. It is however consistent with the tone of the rest of the essay: throughout, Hooykaas does not hide the fact that in his view overconfidence in human reasoning and *a priori* conceptions of reality can hinder productive inquiry into nature. Contingent reality can only be productively researched with humility and proper respect for the empirical facts. Overconfidence in theorizing – especially the Enlightenment glorification of Reason – leads to distortions, resulting in the bounding of science. Considering this, it is easily understood why Hooykaas commends Lavoisier for trying to save the French Academy of the Sciences but criticizes his confidence in human reasoning when it leads him to wrong conclusions.

This is a clear example of how what we can call Hooykaas' philosophy of science – his explicit or implicit views on what science is – informs his historiography of science. For him, science is the interplay between facts (given by nature yet subject to our mode of interpreting them), faith (*a priori* held conceptions of reality, including notions of unity, order, simplicity, and harmony) and fictions (theories, working hypotheses, and conceptualizations produced by the scientist's creative imagination). "When we look at science," Hooykaas submits in a later publication, "we find that fact, faith and fiction are always present together, though in proportions which vary greatly."[33] Although knowledge increases, these three ingredients are always present in both present and past science.

For Hooykaas, the greatest threat to science occurs when the empirical facts are not allowed to regulate human reasoning. Whether it was Marxist scientists or Dutch theologians – Hooykaas self-identified as an "orthodox Protestant"[34] – Hooykaas criticized contemporaries of his if they tried to reduce nature's complexity to fit neatly into their preconceived ideas about reality. For example, in 1949 the so-called Reformed Ecumenical Synod met to discuss the question of evolution, admitting that while the Bible is not a scientific textbook, nevertheless, Genesis 1 does not allow for any theory of evolution. Hooykaas reacted very critically, noting that "the natural scientist is told repeatedly, that he ought to bow before the Word of God, but in fact one sometimes forgets that [...] the theologian has to bow before God's revelation in creation."[35]

He frequently found himself embroiled in "friendly clashes" with his fellow Reformed countrymen.[36] As historian of science Frans van Lunteren explains, Hooykaas found support in the examples of the past:

> devoted Christians for whom the study of God's creation was an act of worship and who, unfettered as they were by the systems and prejudices of their times, were open to the unexpected and simply followed wherever the facts led them. He repeatedly refers to the examples of Kepler, Boyle, Newton, and, above all, Pascal.[37]

Truly, in the words of Arie Leegwater, "if there is one hero that shines through Hooykaas' work, it is Pascal."[38] Time and again Hooykaas calls upon Pascal to testify against overconfidence and pride, like in the essay on Lavoisier. In Pascal, Hooykaas finds an ally against the "authorities of antiquity" and "clerical dignitaries" who try to impose themselves onto nature: "nature exists independently of the opinions humans may form of her" and "their pronunciations cannot alter established facts."[39] In Pascal's time, the lowly pump makers accepted the vacuum as fact; the theorists and philosophers, including René Descartes (1596-1650), tried to wrangle the experimental results into their systems of thought. Just because we do not understand how a vacuum can exist, it nevertheless exists and we must accept it. Hooykaas, thus, approvingly quotes Pascal: "Simple Artisans were capable of pronouncing guilty of error all those great men whom we call Philosophers."[40]

This is, in a nutshell, Hooykaas' philosophy of science: the interplay of fact, faith and fiction; with an utmost respect for facts, requiring a humble attitude towards reality. His insistence on the importance of humbly accepting the empirical facts allowed him to emphasize those historical characters like Boyle and Pascal who, according to Hooykaas, showcased this attitude abundantly. It allowed him to call out aspects in the work of Lavoisier and others which he interpreted as relying too heavily on human reasoning. It also allowed him to approvingly quote someone like atheist agnostic Thomas Huxley (1825-1895): "Sit down before fact like a little child, and be prepared to give up every preconceived notion, follow humbly wherever and to whatever abyss Nature leads or you shall learn nothing."[41] Wherever he found a so-called humble attitude towards empirical reality, Hooykaas praised it. Whereas the heroes in the eyes of Whig-positivist historians were those genius scientists that had freed themselves from theological and philosophical speculation in order to contribute to what is "true" and "positive" scientific knowledge today, Hooykaas' heroes were those with the right humble attitude regardless of whether they contributed to today's body of knowledge or not. When Thomas Kuhn (1922-1996) and other historians from the late 1950s and onwards started emphasizing the philosophy of science like never before, differences between their views and those of Hooykaas would again result in different perspectives on who were considered "heroes" (and why) and who were not.

## Different philosophies of science, different heroes

One historian of science who influenced Kuhn was the Russian born Alexandre Koyré (1892-1964) who, together with Duhem, Metzger, Dijksterhuis, Hooykaas and others, was among the new generation of historians of science. Like the others, he protested the Whig-positivist insistence that science had developed through a series of discoveries of how reality actually works despite the hindering philosophical or theological speculation of the day. Instead, Koyré argued, science is always linked to philosophical theory; an experiment does not lay bare reality as it actually is, it only confirms the theory on which it is based. For example, Koyré downplayed Galileo Galilei's (1564-1642) famous gravitation experiments in that they merely proved the new theory that he had come up with in his mind – Koyré even doubted whether Galileo had actually conducted them. More important was Galileo's shift from Aristotelianism to Neo-Platonic theory which approached reality mathematically, allowing for precision and calculation. On Koyré's account, the scientific revolution" had largely been an intellectual revolution: the "experimental method" as well as technological and social transformations were a result of a change in intellectual attitude – realized by some of "humanity's greatest geniuses" like Galileo and Descartes – not the other way around.[42] He writes:

> It is [...] impossible, as I believe, to give a sociological explanation of the birth of scientific thought, or of the appearance of great geniuses who revolutionized the development of scientific thought – Syracuse does not explain Archimedes, no more than Padua or Florence explains Galileo.[43]

Koyré's intellectual approach, focused on the inner workings of science, was adopted and developed further by historians who sought to distance themselves from Marxist historians who argued that economic and social structures, such as commercial capitalism, determined the shape and development of science. It was no accident that his internal approach to the history of science found "fertile soil" in Cold War America, home to historian Thomas Kuhn.[44]

Drawing inspiration from Koyré, Kuhn started arguing that "normal science" operates in an already established explanatory framework, whether explicitly articulated or not. Scientists' observations are therefore necessarily theory-laden; the same data can yield a different view of reality depending on the operating theory. Transformation in scientific theories, then, does not so much rely on new observations, but more on the reorganization of already existing data along the lines of a newly constructed explanatory framework, or "paradigm."[45] For Kuhn, a clear example of this was the introduction of Copernican theory, that is, the new set of astronomical ideas that replaced geocentrism with heliocentrism. Kuhn pointed out that the publication of Copernicus' *De Revolutionibus* in 1543 was pivotal in this process. In short, Kuhn made the case that new astronomical ideas – that is, a development internal to astronomy – had enormous consequences for the development of astronomy, science generally and matters of philosophy, religion, and Western society at large.[46]

The Koyré-Kuhn line became to be broadly known as the postpositivist approach *par excellence* and grew in dominance from the end of the 1950s onwards. Although Hooykaas had positioned himself from the beginning of his career as an anti-positivist, he did not submit to the postpositivist philosophy of science. Similar to Koyré and Kuhn, he made note of the importance of the theological and philosophical (specifically epistemological) presuppositions of the historical characters he studied. However, Hooykaas did not problematize the act of perceiving empirical data in the way postpositivist historians and philosophers did. For him – as I summarized in the previous section – modern science "acknowledges no authorities (however great they may be) except the authority of nature" and "in case of conflict between [...] rational expectations and [...] discoveries by observation, the investigator's reason must adapt itself to the data provided by nature."[47]

So, while a postpositivist historiography – focused on the internal, theoretical mechanisms of science – gained traction, in Hooykaas' historiography themes like experience over reason and empirical facts over theory shone bright. For example, he took seriously Edgar Zilsel's (1891-1944) so-called Marxist thesis that the origins of early modern science had much to do with the growth of artisan groups, skilled craftsmen with a preference for experimentation over and against rational theorizing. These engineers, metallurgists and others manual labourers freely experimented and did not pay attention to past or present authorities, whether philosophers, poets, or theologians. The Renaissance humanists "looked down upon the mechanical arts and those who cultivated them, the 'mechanical' workers," Hooykaas argued.[48] Much like the alchemists in his earlier studies, Hooykaas found this group of artisans to be wrongly neglected.

Perhaps the group of ordinary people that drew most of his attention were the early-modern Portuguese sailors. Beginning in 1960, Hooykaas started arguing that the Portuguese Voyages of Discovery, conducted by uneducated marines helped pave the way for a shift in worldview by questioning the Ancients' authority. Writing three years after the publication of Kuhn's *The Copernican Revolution*, Hooykaas writes:

> The Portuguese are the people who first of all opened new vistas for Europe in a concrete manner. Not the Copernican world picture was the starting-point of the scientific revolution, but the hard fact, discovered by the intrepid Portuguese seafarers, that the habitable earth was much greater than ancient and medieval philosophy had deemed possible.[49]

The Portuguese quickly realized, through their travels, that the Greeks had wrongly theorized that the equator would be too hot to be inhabitable. Simple observations proved the ancient theories by the likes of Ptolemy to be false, showing that, in many ways, experience was a better judge than reason. "These pioneers," Hooykaas would later write, "who were not hindered by learned prejudices, did not make their decisions whether a certain fact was true by arguments *pro et contra*: for them observation was enough, and facts must be accepted in spite of any apparent 'absurdity.'"[50] Compared to this newly acquired experiential data – whether it was through the Portuguese voyages that "sent a wave of wonder throughout Europe," or a new star in 1572 that "caused not only great surprise but even great anxiety" – the publication of Copernicus' *De Revolutionibus* was "just a ripple on the surface."[51]

Of the astronomers it was Kepler who Hooykaas held in greatest esteem. Copernicus and Galileo had held to an *a priori* conception of the heavenly movements as circular and uniform; full-blown heliocentrism was yet to be empirically demonstrated. Galileo merely inferenced it by analogy based on the observations of four moons orbiting Jupiter and Venus orbiting the Sun. According to Hooykaas, Kepler was no different until he chose to favor the empirical data over reasoning. Kepler realized, due to an eight-minute discrepancy between calculation and observation of the Mars orbit, that the heavenly movements were rather elliptical and erratic. For Hooykaas, it was Kepler, "a lonely man submitted to facts...[who] broke away from a tradition of two thousand years," [52] not Copernicus or Galileo.

**Contextualization: relativizing science?**

The move towards contextualization that was initiated by the first generation of historians of science in the early twentieth century was also a move away from the strong hero language that the Whig-positivists had used. By not taking the present, and its supposed certain knowledge, as definitive judge over history, these historians focused instead on the philosophical and theological ideas that informed the science of a certain period. When they wrote about individuals, they did so not by pointing out the way scientists in the past had contributed to "true scientific knowledge" but the way their science was put together in terms of their ontological and epistemological assumptions.

However, writing history in this way meant inadvertently answering the question: what is science? Although Hooykaas, Koyré, Dijksterhuis, and others from this first generation of historians of science had started a move towards contextualization, they maintained a normative standard about its nature. Koyré thus argued that Galileo's Neo-Platonic approach to science was essentially correct.[53] In Hooykaas' historiography, this tension is very clear. As he writes in the introduction to his dissertation:

> we will have to acknowledge that ideas that now seem foolish, in their time expressed the truth [...] we do not deny the advancement of science, but the path leading to the Tree of Knowledge is winding and often turns on itself [...] this makes us less triumphant and humbler with respect to our ancestors.[54]

Hooykaas despises the triumphalism of the Whig-positivist historians *and* he maintains that science progresses. Although the path to the Tree of Knowledge is winding, there is a Tree of Knowledge.[55]

From the 1960s onward historians of science became increasingly aware that further contextualization meant not only taking into account the philosophical and theological context but also socio-cultural dimensions. Additionally, unlike many of the first generation of historians, they did not view science solely as a force for good. For many, notions like progress had to be discarded to properly contextualize. This realization coincided with an increasing influx of historians without a natural science background in the field. An important example of someone who enabled the contextualist approach is sociologist of knowledge David Bloor (b. 1942). He argued in a book published in 1976 that, to properly contextualize, it is pertinent for a researcher to pragmatically drop such categories as "good" or "bad" science, "true" or "false" knowledge, because whatever science is, it is first and foremost a social phenomenon, an activity like many other activities in society. Scientific knowledge is whatever a scientific community regards as knowledge.[56] Eradicating the aforementioned categories, sociologists, and philosophers who studied laboratory settings from an observer perspective concluded that scientists are not engaging in a special cognitive activity nor are they consistently employing a scientific method. Rather, scientists are pragmatically and contingently occupied with their surroundings, materials, colleagues and machines, using practical reason to solve problems like people in any other setting.[57]

One well-known example in the community of science historians of a work that took some of these new ideas seriously was Steven Shapin and Simon Schaffer's *Leviathan and the Air-Pump: Hobbes, Boyle and the Experimental Life* (1985). Discarding the present appreciation for arriving at scientific truth by way of the "experimental method," Shapin and Schaffer do not describe Boyle as the champion of his mechanical instrument, the air-pump, and empiricism in general. Instead, they position themselves in their book as observers of the debate between Boyle and Hobbes, between two respected methods of knowledge creation, and describe why Boyle's experimentalism gained more credibility than Hobbes' natural philosophy. They argue that the acceptance of the "experimental life" within Royal Society circles had much to do with promoting peace after the civil wars in seventeenth-century England. In short, Boyle's experimental method did not gain traction because it was correct, it gained traction for political reasons.[58]

The dissipation of the belief that science is a distinct activity did not sit well with some historians of science, including Hooykaas. Kuhn, too, although he changed his mind throughout his career and several views on his work exist, can be interpreted to say that science is a special, cumulative activity.[59] For Kuhn, science is unique in its cumulative puzzle-solving abilities, as represented by the succession of paradigms. Although paradigms are incommensurable, comparisons are possible on account of such qualities as simplicity, accuracy and scope. However, what happens is that popular science textbooks, being written in post-revolutionary times, highlight the science and scientists who have contributed to the current normal science. "As pedagogy this technique of presentation is unexceptionable," Kuhn notes, "but when combined with the generally unhistorical air of science writing and with the occasional systematic misconstructions" the idea that "science has reached its present state by a series of individual discoveries and inventions" becomes dominant, "but that is not the way a science develops."[60]

Hooykaas, in an early 1980s article, argues that "among the cultural achievements of mankind, science is characterized by its strongly cumulative character." The subject matter of the historiography of science differs fundamentally from that of the historiography of art, for example:

> A historian of art will not deem van Gogh's painting 'superior' to that of Rembrandt or modern architecture 'better' than that of ancient Egypt. But a historian of science cannot deny that Newton's mechanics is 'better' than that of Aristotle, and Lavoisier's chemistry more conformable to nature than that of Stahl, and he will readily agree that Mendeléev's

> chemistry is superior to that of Lavoisier and that modern geology is superior to that of the end of the 18th century.[61]

However, Hooykaas too remained ever alert for the dangers of focusing on precursors:

> i.e. people who are praiseworthy because more or less vaguely, intuitively, and prophetically they had some foreknowledge of more modern achievements. In many cases this term is well-meant [...] But a precursor is a forerunner, a herald announcing the arrival of the king; he does not exist in his own right but for the sake of him whose 'advent' he announces. In historiography of science this term is meant as praise [...] but in reality it diminishes the status of a person [...] he would be more honoured if he were measured not by what came after him, but received praise for his own achievements in the context of his own epoch.[62]

So when Hooykaas states that "worshipping heroes is as undesirable as disdaining them,"[63] it is the former because it reduces history's complexity and it is the latter because it does not even do justice to the individual's accomplishments.

## Contextualization: relativizing the individual?

Further contextualization not only raised questions about the special nature of science, but also about the role of the individual in science. This was already evident among postpositivist writers. As philosopher of science Michael Polanyi remarked:

> [T]he example of great scientists is the light which guides all workers in science...[but] we must guard against being blinded by it. There has been too much talk about the flash of discovery and this has tended to obscure the fact that discoveries, however great, can only give effect to some intrinsic potentiality of the intellectual situation in which scientists find themselves.[64]

To put it in blunt terms: is it not the case that – because of broader developments – if not Copernicus, another astronomical thinker would have come up with the same theory around the same time, hence relativizing the accomplishment of the individual? For Kuhn, "the extent of the innovation that any individual can produce is necessarily limited, for each individual must employ in his research the tools that he acquires from a traditional education, and he cannot in his own lifetime replace them all." For him, Copernicus did, however, leave his fingerprints all over *De Revolutionibus*. Copernicus' "eye so absorbed with geometrical harmony" at the same time allowed him to conclude correctly that earth must be moving, conclude wrongly that heavenly movements are circular and uniform and disregard the nonastronomical implications of his theory, helping him to accept it.[65] Still, even in this frame Copernicus' absorbed eye is viewed as a product of his specialist training.

An increasing emphasis on the socio-cultural nature of science added to a further contextualization of individual scientists' achievements. In addition to, and somewhat different from the Kuhnian postpositivist approach, historiography started focusing on several different aspects, including scientific practices, material conditions, tools and objects, linguistic practices, locations and institutionalization dynamics. A radical version of this development is Bruno Latour's (1947-2022) idea to stop distinguishing between human and non-human entities (e.g. a microscope) because both as social actors are part of a complex web of relations.[66] Although Latour's arguments have remained controversial among science historians, the study of objects has gained traction as shown by the publication of Lorraine Daston's edited volume *Biographies of Scientific Objects*.[67]

Hooykaas' philosophy of science, undergirded by his belief that God created a contingent reality, entailed that science is a fundamentally human endeavour and that individuals cannot be merely determined by their environment. In 1983, he wrote that

> neither these general external influences nor the internal development of science should make us blind to the fact that, in various degrees, scientists put the stamp of their unique personality on their work [...] scientists are more than mere products of philosophical, religious, social and economic 'influences'.[68]

In the 1976 Gifford Lectures, given at the University of St Andrews, Hooykaas had already briefly touched on the methodological issue that he, after Pascal, quipped "Cleopatra's nose." Had her nose have been shorter, perhaps Mark Anthony would not have fallen in love with the Egyptian queen and history would have run a different course. Hooykaas asked: is history determined by "fixed socio-historical laws" or rather "capriciously determined by contingencies?"[69] He pointed out that the old historiography busied itself with individuals, but that in his time the attention has gone to society at large and economic conditions. He noted:

> In our day the pendulum has swung back and hero worship, though not extinct, is unfashionable. The 'forerunners' of a development are not now seen as subservient to the great, but rather as proofs that the time was ripe, that the great discoveries had to come and that they did not originate in a 'catastrophic' way as by a flash of lightning.[70]

Having been one of the first historians of science to vehemently argue against the historiography of genius, Hooykaas felt the force of the new perspective. He agreed that some scientific novelties occur simply because "their time has come" or are "in the air" like Newton's law of gravitation. Likewise, both Beeckman and Galileo were separately working on the law of falling bodies. So, although Newton's and Galileo's "powers of synthesis...[and] their mathematical and systematic talents" distinguished them from others, "they were [all] explorers on the same beach."[71] Hooykaas, however, also found instances where seemingly contingent circumstances played a vital role in the development of science. For example, he noticed how Dalton's high esteem of Newton caused him to reject Avogadro's theories, therefore delaying the adoption of the latter. This example showed Hooykaas that "as well as activating the growth of science, great scientists have often had a stifling influence upon it. However much modern science can still profit from their work, the Newtons, Lyells, and Darwins become a dead weight when they are taken as unquestionable authorities."[72]

So, according to Hooykaas, both the greater whole and the individual are important to take into consideration. He does not provide an elaborate philosophy of science to flesh this out. Here, too, he explicitely resists what he deems to be the temptation to systematize:

> For me, to be a historian means to yield all my powers and imagination to the enterprise of reliving the past, and entering into the minds of those who went before us. I see it as neither possible nor pertinent to draw from history the kinds of lesson that can be expressed in laws and systems. To have lived with these people, to have come to recognize ourselves to a certain extent in their human strengths and weaknesses – for me this is the reward of the study of history.[73]

## Concluding thoughts

The nineteenth-century image of the genius scientist who unearthed certain, "positive" knowledge despite prevailing hindering religious and philosophical speculation was discarded by the first generation of professional science historians in the first half the twentieth century. These historians contextualized individual scientists by studying the relevant theological and philosophical (specifically epistemological) ideas. This did not stop historians from writing more positively about certain historical characters than others. Although wanting to approach history on its own terms, they too had their normative standards. The normative claim Hooykaas made was the precedence of experience over reason, a presupposition fueled by the required humility in the face of contingent reality created by God.

Hooykaas praised the attitude of humility, a radical openness to reality wherever he found it, whether it was Kepler, Pascal, Thomas Huxley, Bakhuis Roozeboom, or the neglected alchemists and Portuguese sailors. Together with his contemporaries like Dijksterhuis and Koyré, Hooykaas shared with the Whig-positivists the general belief that science is a special, cumulative activity. But because he prioritized the virtue of humility he vehemently argued against the idea of the genius-scientist hero.

Slowly but surely further contextualization weakened the hold of philosophy over historiography from the 1960s onward. Established historians – postpositivsts and Hooykaas among them – protested by attempting to hold on to the cumulative and therefore special nature of science. Likewise, as the context grew the role of the individual was relativized. Because of these developments, strongly positive (or negative) evaluations of individuals in professional historiographical works diminished. In the early 1980s, Hooykaas repeated what he had already stated in his 1933 dissertation: the most important pitfalls in historiography are nationalism, hero-worship, and a general sense of superiority towards previous generations.[74] For historians, including the ones I have mentioned – Shapin, Schaffer, and Daston – who (partially) follow the insights of Bloor and Latour this was not enough, to the point that the historiography of science has today "for the most part [become] a non-evaluative discipline."[75] Describing over judging has become so important that, in some historians' estimation, this almost amounts to a return to Leopold von Ranke's naive nineteenth-century dictum about the goal of historiography, that is, the idea of "*bloss zeigen wie es eigentlich gewesen*": recording of history without any evaluation or story to tell.[76] Naturally, pure description (if that is even possible) results in histories without heroes, obstacles, solutions, nothing to appreciate and nothing to learn from.

One relevant countertrend to the sociological turn in the historiography of science has been biography. As historian of science Helge Kragh wrote in 2015: it is no longer "necessary to defend the art of biography as an essential part of history of science."[77] Biography both localizes and humanizes the history of science. It is at the human level where both the theoretical background understanding or social phenomena are manifested. It is in this way that biography can be a "literary lens" through which to study science.[78] Depending on whether the biographer treats the biographee's science as a special activity, criticism from sociologically-minded historians will vary. Another potential pitfall of the biographical method has been identified by science historian Mott Greene: the biographer needs to work hard to avoid the hero trope. Biography entails story and stories have beginnings and endings, obstacles and people overcoming said obstacles. Especially biographers of creatives like Einstein need to tread carefully if they want to avoid turning their writing into a "hero's quest."[79] But as a biography on twentieth-century physicist Fritz London – a relatively unknown figure – shows, a narrative about an individual doing, somewhat boring, "normal science,"[80] can be "higly interesting" and avoid both hero worship and radical contextualization.[81] For example, the biographer decided that philosophical and cultural factors were important, whereas political and economic factors were not.

Either way, almost only biographies become best sellers with the wider public. A 1991 biography of Darwin sold more than 100,000 copies.[82] There are numerous other examples that show that the public at large tends to read, of all the works science historians produce, biographies of scientists. This same is true for scientists themselves. Twenty of the fifty recipients of the Dexter and Edelstein Awards – awarded by the American Chemical Society for exceptional contributions to the history of chemistry – in the period from 1956 to 2006, had written biographical works. Six of these were about Antoine Lavoisier.[83] In fact, historian of science Bernadette Bensaude-Vincent wrote in 1996 that "the image of Lavoisier as the founding father of modern chemistry still reigns supreme in the collective memory of professional chemists, at least in France."[84] The case of Lavoisier serves as a reminder that popular and professional historiographical memories do not always align. In the Netherlands, Dutch chemist Hendrik Bakhuis Roozeboom is commemorated through a medal bearing his name, awarded by the KNAW to scholars who have contributed substantially to the phase theory of elements. Since 1916 it has been awarded sixteen times to researchers worldwide.[85] The case of Bakhuis Roozeboom – and Hooykaas' commemoration – serves as a reminder that professional science historians do not always agree either: who deserves their history writing and why?

## Primary sources

Reijer Hooykaas Papers. Noord-Hollands Archief, Haarlem, Netherlands.

## Notes

1 Reijer Hooykaas, "Pitfalls in the Historiography of Geological Science," *Histoire et Nature* 19-20 (1981-1982), 21–34, on 26. I would like to thank Ab Flipse, Gijsbert van den Brink, Matthias Smalbrugge and Ad Maas for commenting on earlier versions of this text. Any remaining mistakes are mine.

2 Reijer Hooykaas, "Hendrik Willem Bakhuis Roozeboom (1854-1907): Grondlegger der phasenleer," *Geloof en Wetenschap* 53 (1955), 69–77, on 69. In this paper I will translate all non-English quotes to English. When the original phrase sheds additional light on the conveyed meaning, I will include it in-text.

3 Andreas Smits mentioned this during a lecture at the Technische Hoogeschool Delft, as published in newspaper *De Standaard*, February 18, 1907; H. R. Kruyt, "Eeuwherdenking H. W. Bakhuis

Roozeboom," *Chemisch Weekblad* 50 (1954), 749–753, on 752; J. Fokkens, "H. W. Bakhuis Roozeboom," *Geloof en Wetenschap* 47 (1949), 115–122, on 122.

4 I quote here from the contribution by Ad Maas and Louise Lagarde in this volume.

5 The other two professional associations were the *Koninklijke Nederlandse Chemische Vereniging* (Royal Netherlands Chemical Society) and the *Genootschap voor Natuur-, Genees- en Heelkunde* (Society for the Advancement of Science, Medicine and Surgery). For a report on these memorial activities see *Chemisch Weekblad* 50, no 44 (1954), a special edition dedicated to the commemoration of Bakhuis Roozeboom.

6 According to letters from J. F. Koksma and T. van der Linden to Reijer Hooykaas, August 18 and September 27, 1954, in the Hooykaas Papers, Noord-Hollands Archief, Haarlem, Netherlands (hereafter HP), Hooykaas had been invited to be part of the organizing committee of the memorial activities co-organized by the KNAW, but he had no formal role during the days themselves. A note on my usage of the archive: unfortunately, due to current inventory and reorganization efforts of the Hooykaas Papers by the Noord-Hollands Archief, no specific references could be provided.

7 The speech was later published as Reijer Hooykaas, "Hendrik Willem Bakhuis Roozeboom (1854-1907): Grondlegger der phasenleer," *Geloof en Wetenschap* 53 (1955): 69–77, quotations from 69–71. A shortened version was published in newspaper *Trouw*, October 23, 1954.

8 Ab Flipse, "Reijer Hooykaas (1906-1994)," *Studium* 6 (2013), 287–291.

9 Leen Dorsman, "Ernst Cohen (1869–1944): chemicus, bestuurder, geschiedschrijver," *Studium* 6 (2013), 237–240. See also H. A. M. Snelders, *De geschiedenis van de scheikunde in Nederland: Van alchemie tot chemie en chemische industrie rond 1900* (Delft: Delftse Universitaire Pers, 1993), 5.

10 For the particular way this played out in the historiography of chemistry, see John G. McEvoy, *The Historiography of the Chemical Revolution: Patterns of Interpretation in the History of Science* (London: Pickering & Chatto, 2010). For an analysis on the close connection between the origins of genius, science and historiography, see Simon Schaffer, "Discoveries and the End of Natural Philosophy," *Social Studies of Science* 16 (1986), 387–420.

11 Bernadette Bensaude-Vincent, "A Founder Myth in the History of Science? The Lavoisier Case," in *Functions and Uses of Disciplinary Histories*, ed. Loren Graham, Wolf Lepenies and Peter Weingart (Dordrecht: D. Reidel, 1983), 53–78, on 53.

12 McEvoy, *The Historiography of the Chemical Revolution*, 23–34.

13 David Baneke, "De rol van een genootschap. Honderd jaar Gewina in het wetenschapshistorische landschap," *Studium* 6 (2013), 132–148, on 135.

14 Klaas van Berkel, "Natuurwetenschap en cultureel nationalisme in negentiende-eeuws Nederland," in *Citaten uit het boek der natuur. Opstellen over Nederlandse wetenschapsgeschiedenis*, ed. Klaas van Berkel (Amsterdam: Bert Bakker, 1998), 221–239, on 239.

15 Eduard Jan Dijksterhuis, *Val en worp: een bijdrage tot de geschiedenis der mechanica van Aristoteles tot Newton* (Groningen: P. Noordhoff, 1924). Five years before, the cultural historian Johan Huizinga had written a masterpiece about the Late Middle Ages to try to capture the feel of the age. He rejected reducing this period to merely a precursor to the Renaissance: *Autumn of the Middle Ages*, trans. Rodney J. Payton and Ulrich Mammitzsch (Chicago: Chicago University Press, 2020 [1919]). Hooykaas refers to neither in his earliest works.

16 Reijer Hooykaas, *Het begrip element in zijn historisch-wijsgeerige ontwikkeling* (Utrecht: Schotanus & Jens, 1933).

17 Cf. H. R. Kruyt, "Afscheidscollege," *Chemisch Weekblad* 42 (1946), 264–270, on 267.

18 Hooykaas, *Het begrip element*, 12–14.

19 H. Floris Cohen, "Eloge: Reijer Hooykaas, 1 August 1906–4 January 1994," *Isis* 89 (1998), 181–184, on 181.

20 Hooykaas, *Het begrip element*, 11.

21 Bernadette Bensaude-Vincent, "Chemistry in the French Tradition of the Philosophy of Science: Duhem, Meyerson, Metzger and Bachelard," *Studies in History and Philosophy of Science* 36 (2005), 627–649, on 645. In his dissertation, Hooykaas refers to Pierre Duhem, *Le mixte et la combination chimique* (Paris: Fayard, 1902); Hélène Metzger, *Les doctrines chimiques en France du début du XVIIIe à la fin du XVIIIe siècle* (Paris: Presses universitaires de France, 1923); *Newton, Stahl, Boerhaave et la doctrine chimique* (Paris: Alcan, 1930); and Émile Meyerson, *Identité et realité*, 3rd ed. (Paris: Alcan, 1926).

22 Bernadette Vincent-Bensaude, "Hélène Metzger's 'La Chimie': A Popular Treatise," *History of Science* 25 (1987), 71–84, on 73–74.

23 George Sarton, review of *Les doctrines chimiques en France du début du XVIIe à la fin du XVIIIe siècle*, by Hélène Metzger, *Isis* 6 (1924), 57–64, on 63–64. As I have done with Dutch quotes in this paper, I have translated the original French here into English.

24 Sarton did not extensively evaluate Hooykaas' dissertation although he did make note of it, as he wrote a few lines accompanying its entry for *Isis*'s Critical Bibliography: "Very elaborate study of the evolution of the concept of 'element.' The thesis is written in Dutch but is followed with a French summary of 4 pages." George Sarton, bibliography entry for *Het begrip element in zijn historisch-wijsgeerige ontwikkeling*, by Reijer Hooykaas, Forty-Second Critical Bibliography of the History and Philosophy of Science and of the History of Civilization, *Isis* 23 (1935), 586.

25 Sarton, review, 64. On Sarton see Geert Somsen, "George Sarton (1884-1956)," *Studium* 6 (2013), 259–262.

26 John Hedley Brooke and Michael Hunter, foreword to *Robert Boyle: A Study in Science and Christian Belief*, by Reijer Hooykaas, trans. H. van Dyke (Lanham: University Press of America, 1997), xix (italics original), originally published as *Robert Boyle: een studie over natuurwetenschap en christendom* (Loosduinen: Kleijwegt, 1942). "Pascal: zijn wetenschap en religie," *Orgaan van de Christelijke Vereeniging van Natuur- en Geneeskundigen in Nederland* (1939), 147–178, received two translations, the latest of which can be found as "Pascal: His Science and His Religion," in *Tractrix: Yearbook for the History of Science, Medicine, Technology and Mathematics* 1 (1990), 115–139 (trans. H. Floris Cohen). In this text I will refer to this English edition.

27 Hooykaas, *Robert Boyle*, xxii–xxiii.

28 Reijer Hooykaas, *De chemische omwenteling: Lavoisier* (Arnhem: Van Loghum Slaterus, 1952). This is somewhat of a popularizing essay: although it quotes both Lavoisier and others, no footnotes are provided.

29 The fact Hooykaas discusses so many different activities was perhaps inspired by Dijksterhuis' biography of Simon Stevin that did the same thing: *Simon Stevin: Science in the Netherlands around 1600* (The Hague: Martinus Nijhoff, 1970 [1943]).

30 Hooykaas, *De chemische omwenteling*, 31–32.

31 Ibid., 20, 31.

32 Ibid., 60.

33 Reijer Hooykaas, *Fact, Faith and Fiction in the Development of Science: The Gifford Lectures Given in the University of St Andrews 1976*, with a foreword by J. Christiaan Boudri, H. Floris Cohen and Valerie MacKay (Dordrecht: Kluwer Academic, 1999), 14.

34 Reijer Hooykaas to Robert E. Vander Vennen, 17 November 1965, HP.

35 Reijer Hooykaas, "Dominees en evolutie," *Bezinning* 5 (1950), 74–88, on 76.

36 He uses the phrase "friendly clash" to describe his encounter with Vrije Universiteit theologians at the 1948 International Calvinistic Conference hosted in Amsterdam: Reijer Hooykaas to Douglas Johnson, 5 April 1987, HP.

37 Frans van Lunteren, "R. Hooykaas, *Natural Law and Divine Miracle*: Geology and Christianity," *Isis* 109 (2018), 122–126, on 124.

38 Arie Leegwater, "Reijer Hooykaas (1906-1994): A Modern Advocate for *Philosophia Libera*," *Perspectives on Science and Christian Faith* 48 (1996), 98–103, on 99. Incidentally, after the war Dijksterhuis too became an admirer of Pascal, especially of his emphasis on mathematical and disciplined thinking: "De betekenis van de wis- en natuurkunde voor het leven en denken van Blaise Pascal," in *Clio's stiefkind*, ed. Klaas van Berkel (Amsterdam: Bert Bakker, 1990). For more information on how Pascal influenced Hooykaas, see my "Pascal in de wetenschapshistoriografie van Reijer Hooykaas," *Sophie* 13, no. 3 (2023), 28–31.

39 Hooykaas, "Pascal," 123.

40 Ibid., 127.

41 Reijer Hooykaas, *Natural Law and Divine Miracle: A Historical-Critical Study of the Principle of Uniformity in Geology, Biology, and Theology* (Leiden: Brill, 1959), 210.

42 Alexandre Koyré, *Études galiléennes* (Paris: Hermann, 1966 [1939]), 15.

43 Alexandre Koyré, *Études d'histoire de la pensée philosophique* (Paris: Gallimard, 1971 [1961]), 223–224.

44 Alexandre Koyré, "Galileo and Plato," *Journal of the History of Ideas* 4 (1943), 400–428; McEvoy, *The Historiography of the Chemical Revolution*, 85.

45 Thomas S. Kuhn, *The Structure of Scientific Revolutions*, 4th edition (Chicago: Chicago University Press, 2012 [1962]).

46 Thomas S. Kuhn, *The Copernican Revolution: Planetary Astronomy in the Development of Western Thought* (Cambridge, MA: Harvard University Press, 1957).

47 Reijer Hooykaas, "The Rise of Modern Science: When and Why?" *The British Journal for the History of Science* 20 (1987), 453–473, on 455.

48 Ibid., 461. Edgar Zilsel had already written about the concept of genius and specifically about it in relation to the Renaissance years before in, e.g., his *Die Geniereligion: Ein kritischer Versuch über das moderne Persönlichkeitsideal, mit einer historischen Begründung* (Frankfurt: Suhrkamp, 1990 [1918]).

49 Reijer Hooykaas, "The History of Portuguese Culture," *Free University Quarterly* 7 (1960), 204–211, on 211.

50 See also Hooykaas, "The Rise of Modern Science," 459.

51 Reijer Hooykaas, "The Impact of the Copernican Transformation," in *The "Conflict Thesis" and Cosmology*, Science and Belief: From Copernicus to Darwin, block 1 unit 2 (Milton Keynes: Open University Press, 1974), 51–86, on 51.

52 Reijer Hooykaas, *Religion and the Rise of Modern Science* (Vancouver: Regent College Publishing, 2000 [1972]). For an early publication on Kepler see Reijer Hooykaas, "Het hypothesenbegrip van Kepler," *Orgaan van de Christelijke Vereeniging van Natuur- en Geneeskundigen in Nederland* (1939), 38–55.

53 Koyré, *Études galiléennes.*

54 Hooykaas, *Het begrip element*, 12–14.

55 This tension in the first generation of professional historians of science is noted by Bart Karstens. Even though they retained some Whiggish undertones, they were not naively Whiggish: Bart Karstens, "The Peculiar Maturation of the History of Science," in *The Making of the Humanities*, ed. by Rens Bod, Jaap Maat and Thijs Weststeijn (Amsterdam: Amsterdam University Press, 2014), 3: 183–203.

56 David Bloor, *Knowledge and Social Imagery* (Chicago: Chicago University Press, 1991 [1976]).

57 E.g., Bruno Latour and Steve Woolgar, *Laboratory Life: The Construction of Scientific Facts*, 2nd ed. (Princeton, NJ: Princeton University Press, 1986 [1979]); Karin Knorr-Cetina, *The Manufacture of Knowledge: An Essay on the Constructivist and Contextual Nature of Science* (Oxford: Pergamon

Press, 1981); Michael Lynch, *Art and Artifact in Laboratory Science: A Study of Shop Work and Shop Talk in a Research Laboratory* (London: Routledge, 1984).

58 Steven Shapin and Simon Schaffer, *Leviathan and the Air-Pump: Hobbes, Boyle and the Experimental Life* (Princeton, NJ: Princeton University Press, 1985).

59 See for an evaluation of Kuhn's ideas Jouni-Matti Kuukkanen, *Meaning Changes. A Study of Thomas Kuhn's Philosophy* (Saarbrücken: Akademiker Verlag, 2008).

60 Kuhn, *The Structure of Scientific Revolutions*, 139–140.

61 Hooykaas, "Pitfalls in the Historiography of Geological Science," 24.

62 Ibid., 26.

63 Reijer Hooykaas, presentation at a meeting of the *Commissie voor de Geschiedenis der Geologische Wetenschappen*: Geschiedenis Geologie in Nederland, November 7, 1975, HP.

64 Michael Polanyi, "My Time with X-Rays and Crystals," in *Knowing and Being*, ed. Marjorie Grene (London: Routledge, 1969), 97–104, on 97.

65 Kuhn, *The Copernican Revolution*, 183–184.

66 Bruno Latour, *Science in Action: How to Follow Scientists and Engineers through Society* (Cambridge, MA: Harvard University Press, 1987).

67 *Biographies of Scientific Objects*, ed. Lorraine Daston (Chicago: Chicago University Press, 2000).

68 Reijer Hooykaas, "The Concept of 'Element' in Its Historical-Philosophical Development," in *Selected Studies in History of Science* (Coimbra: Universidade Coimbra, 1983), 8.

69 Hooykaas, *Fact, Faith and Fiction in the Development of Science*, 19.

70 Ibid., 337.

71 Ibid., 337.

72 Ibid., 335.

73 Ibid., 341.

74 Hooykaas, "Pitfalls in the Historiography of Geological Science."

75 Bart Karstens, "Pluralism within Parameters: Towards a Mature Evaluative Historiography of Science" (PhD thesis, Universiteit Leiden, 2015), 5.

76 Karstens, "The Peculiar Maturation of the History of Science," 195. See also Jouni-Matti Kuukkanen, "The Missing Narrativist Turn in the Historiography of Science," *History and Theory* 51 (2012), 340–363.

77 Helge Kragh, "On Scientific Biography and Biographies of Scientists," in *Relocating the History of Science: Essays in Honor of Kostas Gavroglu*, ed. Theodore Arabatzis, Jürgen Renn and Ana Simões (Heidelberg: Springer 2015), 269–280, on 269.

78 Thomas L. Hankins, "In Defence of Biography: The Use of Biography in the History of Science," *History of Science* 17 (1979), 1–16, on 14.

79 Mott T. Greene, "Writing Scientific Biography," *Journal of the History of Biology* 40 (2007), 727–759.

80 Kostas Gavroglu, *Fritz London: A Scientific Biography* (Cambridge: Cambridge University Press, 1995), xv.

81 Kragh, "On Scientific Biography and Biographies of Scientists," 270, 275.

82 Adrian Desmond and James Moore, *Darwin* (London: Michael Joseph, 1991). See Thomas Söderqvist, "A New Look at the Genre of Scientific Biography," in *The History and Poetics of Scientific Biography*, ed. by Thomas Söderqvist (Aldershot, UK: Ashgate, 2007), 1–15, on 2.

83 Mary Jo Nye, "Scientific Biography in the History of Chemistry: The Role of Dexter and Edelstein Award Winners in the Last Fifty Years," *Bulletin for the History of Chemistry* 32 (2007), 21–26, on 21.

84 Bernadette Bensaude-Vincent, "Between History and Memory: Centennial and Bicentennial Images of Lavoisier," *Isis* 87 (1996), 481–499, on 482.

85 https://www.knaw.nl/en/funds-and-prizes/bakhuis-roozeboom-medal.

## Bibliography


Baneke, David. "De rol van een genootschap. Honderd jaar Gewina in het wetenschapshistorische landschap." *Studium* 6 (2013): 132–148.

Bensaude-Vincent, Bernadette. "A Founder Myth in the History of Science? The Lavoisier Case." In *Functions and Uses of Disciplinary Histories*, edited by Loren Graham, Wolf Lepenies and Peter Weingart, 33–70. Dordrecht: D. Reidel, 1983.

Bensaude-Vincent, Bernadette. "Hélène Metzger's 'La Chimie': A Popular Treatise." *History of Science* 25 (1987): 71–84.

Bensaude-Vincent, Bernadette. "Between History and Memory: Centennial and Bicentennial Images of Lavoisier." *Isis* 87 (1996): 481–499.

Bensaude-Vincent, Bernadette. "Chemistry in the French Tradition of the Philosophy of Science: Duhem, Meyerson, Metzger and Bachelard." *Studies in History and Philosophy of Science* 36 (2005): 627–649.

Bloor, David. *Knowledge and Social Imagery*. Chicago: Chicago University Press, 1991 [1976].

Brooke, John Hedley, and Michael Hunter. Foreword to *Robert Boyle: A Study in Science and Christian Belief*, by Reijer Hooykaas, vii-xix. Lanham: University Press of America, 1997.

*Chemisch Weekblad* 50, no. 44 (1954).

Cohen, H. Floris. "Eloge: Reijer Hooykaas, 1 August 1906–4 January 1994." *Isis* 89 (1998): 181–184.

Daston, Lorraine, ed. *Biographies of Scientific Objects*. Chicago: Chicago University Press, 2000.

Desmond, Adrian, and James Moore. *Darwin*. London: Michael Joseph, 1991.

Dijksterhuis, E. J. *Val en worp: een bijdrage tot de geschiedenis der mechanica van Aristoteles tot Newton*. Groningen: P. Noordhoff, 1924.

Dijksterhuis, E. J. *Simon Stevin: Science in the Netherlands around 1600*. The Hague: Martinus Nijhoff, 1970 [1943].

Dijksterhuis, E. J. "De betekenis van de wis- en natuurkunde voor het leven en denken van Blaise Pascal." In *Clio's stiefkind*, edited by Klaas van Berkel, 67–101. Amsterdam: Bert Bakker, 1990 [1952].

Dorsman, Leen. "Ernst Cohen (1869–1944): chemicus, bestuurder, geschiedschrijver." *Studium* 6 (2013): 237–240.

Flipse, Ab. "Reijer Hooykaas (1906-1994)." *Studium* 6 (2013): 287–291.

Fokkens, J. "H. W. Bakhuis Roozeboom." *Geloof en Wetenschap* 47 (1949): 115–122.

Gavroglu, Kostas. *Fritz London: A Scientific Biography*. Cambridge: Cambridge University Press, 1995.

Greene, Mott T. "Writing Scientific Biography." *Journal of the History of Biology* 40 (2007): 727–759.

Hankins, Thomas L. "In Defence of Biography: The Use of Biography in the History of Science." *History of Science* 17 (1979): 1–16.

Heeren, Jelmer. "Pascal in de wetenschapshistoriografie van Reijer Hooykaas." *Sophie* 13, no. 3 (2023): 28–31.

Hooykaas, Reijer. *Het begrip element in zijn historisch-wijsgeerige ontwikkeling*. Utrecht: Schotanus & Jens, 1933.

Hooykaas, Reijer. "Het hypothesenbegrip van Kepler." *Orgaan van de Christelijke Vereeniging van Natuur- en Geneeskundigen in Nederland* (1939): 38–55.

Hooykaas, Reijer. "Pascal: His Science and His Religion." Translated by H. Floris Cohen. *Tractrix: Yearbook for the History of Science, Medicine, Technology and Mathematics* 1 (1990 [1939]): 115–139.

Hooykaas, Reijer. *Robert Boyle: A Study in Science and Christian Belief*. With a foreword by John Hedley Brooke and Michael Hunter. Translated by H. van Dyke. Lanham: University Press of America, 1997 [1942].

Hooykaas, Reijer. "Dominees en evolutie." *Bezinning* 5 (1950): 74–88.

Hooykaas, Reijer. *De chemische omwenteling: Lavoisier.* Arnhem: Van Loghum Slaterus, 1952.

Hooykaas, Reijer. "Hendrik Willem Bakhuis Roozeboom (1854-1907): Grondlegger der phasenleer." *Geloof en Wetenschap* 53 (1955): 69–77.

Hooykaas, Reijer. *Natural Law and Divine Miracle: A Historical-Critical Study of the Principle of Uniformity in Geology, Biology, and Theology*. Leiden: Brill, 1959.

Hooykaas, Reijer. "The History of Portuguese Culture." *Free University Quarterly* 7 (1960): 204–211.

Hooykaas, Reijer. *Religion and the Rise of Modern Science.* Vancouver: Regent College Publishing, 2000 [1972].

Hooykaas, Reijer. "The Impact of the Copernican Transformation." In *Science and Belief: From Copernicus to Darwin. The "Conflict Thesis" and Cosmology*, block 1 unit 2, 51–86. Milton Keynes: Open University Press, 1974.

Hooykaas, Reijer. "Pitfalls in the Historiography of Geological Science." *Histoire et Nature* 19-20 (1981-1982): 21–34.

Hooykaas, Reijer. "The Concept of 'Element' in Its Historical-Philosophical Development." In *Selected Studies in History of Science*, 7–8. Coimbra: Universidade Coimbra, 1983.

Hooykaas, Reijer. "The Rise of Modern Science: When and Why?" *The British Journal for the History of Science* 20 (1987): 453–473.

Hooykaas, Reijer. *Fact, Faith and Fiction in the Development of Science: The Gifford Lectures Given in the University of St Andrews 1976.* With a foreword by J. Christiaan Boudri, H. Floris Cohen and Valerie MacKay. Dordrecht: Kluwer Academic, 1999.

Huizinga, Johan. *Autumn of the Middle Ages*. Translated by Rodney J. Payton and Ulrich Mammitzsch. Chicago: Chicago University Press, 2020 [1919].

Karstens, Bart. "The Peculiar Maturation of the History of Science." In *The Modern Humanities*, edited by Rens Bod, Jaap Maat and Thijs Weststeijn, 183–203. Vol. 3 of *The Making of the Humanities.* Amsterdam: Amsterdam University Press, 2014.

Karstens, Bart. "Pluralism within Parameters: Towards a Mature Evaluative Historiography of Science." PhD diss., Universiteit Leiden, 2015.

Knorr-Cetina, Karin. *The Manufacture of Knowledge: An Essay on the Constructivist and Contextual Nature of Science.* Oxford: Pergamon Press, 1981.

Koyré, Alexandre. "Galileo and Plato." *Journal of the History of Ideas* 4 (1943): 400–428.

Koyré, Alexandre. *Études galiléennes.* Paris: Hermann, 1966 [1939].

Koyré, Alexandre. *Études d'histoire de la pensée philosophique.* Paris: Gallimard, 1971 [1961].

Kragh, Helge. "On Scientific Biography and Biographies of Scientists." In *Relocating the History of Science: Essays in Honor of Kostas Gavroglu*, edited by Theodore Arabatzis, Jürgen Renn and Ana Simões, 269–280. Heidelberg: Springer, 2015.

Kruyt, H. R. "Afscheidscollege." *Chemisch Weekblad* 42 (1946): 264–270.

Kruyt, H. R. "Eeuwherdenking H. W. Bakhuis Roozeboom." *Chemisch Weekblad* 50 (1954): 749–753.

Kuhn, Thomas S. *The Copernican Revolution: Planetary Astronomy in the Development of Western Thought.* Cambridge, MA: Harvard University Press, 1957.

Kuhn, Thomas S. *The Structure of Scientific Revolutions.* 4th edition. Chicago: Chicago University Press, 2012 [1962].

Kuukkanen, Jouni-Matti. *Meaning Changes. A Study of Thomas Kuhn's Philosophy.* Saarbrücken: Akademiker Verlag, 2008.

Kuukkanen, Jouni-Matti. "The Missing Narrativist Turn in the Historiography of Science." *History and Theory* 51 (2012): 340–363.

Latour, Bruno, and Steve Woolgar. *Laboratory Life: The Construction of Scientific Facts.* 2nd ed. Princeton, NJ: Princeton University Press, 1986 [1979].

Lynch, Michael. *Art and Artifact in Laboratory Science: A Study of Hop Work and Shop Talk in a Research Laboratory.* London: Routledge, 1984.

Metzger, Hélène. *Les doctrines chimiques en France du début du XVIIIe à la fin du XVIIIe siècle.* Paris: Presses universitaires de France, 1923.

McEvoy, John G. *The Historiography of the Chemical Revolution: Patterns of Interpretation in the History of Science.* London: Pickering & Chatto, 2010.

Nye, Mary Jo. "Scientific Biography in the History of Chemistry: The Role of Dexter and Edelstein Award Winners in the Last Fifty Years." *Bulletin for the History of Chemistry* 32 (2007): 21–26.

Polanyi, Michael. "My Time with X-Rays and Crystals." In *Knowing and Being*, edited by Marjorie Grene, 97–104. London: Routledge, 1969.

Sarton, George. Review of *Les doctrines chimqiues en France du début du XVIIe à la fin du XVIIIe siècle*, by Hélène Metzger. *Isis* 6 (1924), 57–64.

Sarton, George. Bibliography entry for *Het begrip element in zijn historisch-wijsgeerige ontwikkeling*, by Reijer Hooykaas. Forty-Second Critical Bibliography of the History and Philosophy of Science and of the History of Civilization, *Isis* 23 (1935): 586.

Shapin, Steven, and Simon Schaffer. *Leviathan and the Air-Pump: Hobbes, Boyle and the Experimental Life.* Princeton, NJ: Princeton University Press, 1985.

Schaffer, Simon. "Discoveries and the End of Natural Philosophy." *Social Studies of Science* 16 (1986): 387–420.

Snelders, H. A. M. *De geschiedenis van de scheikunde in Nederland: Van alchemie tot chemie en chemische industrie rond 1900.* Delft: Delftse Universitaire Pers, 1993.

Somsen, Geert. "George Sarton (1884-1956)." *Studium* 6 (2013): 259–262.

Söderqvist, Thomas. "A New Look at the Genre of Scientific Biography." In *The History and Poetics of Scientific Biography*, edited by Thomas Söderqvist, 1–15. Aldershot, UK: Ashgate, 2007.

Van Berkel, Klaas. "Natuurwetenschap en cultureel nationalisme in negentiende-eeuws Nederland." In *Citaten uit het boek der natuur. Opstellen over Nederlandse wetenschapsgeschiedenis*, edited by Klaas van Berkel, 221–239. Amsterdam: Bert Bakker, 1998.

Van Lunteren, Frans. "R. Hooykaas, *Natural Law and Divine Miracle*: Geology and Christianity." *Isis* 109 (2018): 122–126.

Zilsel, Edgar. *Die Geniereligion: Ein kritischer Versuch über das moderne Persönlichkeitsideal, mit einer historischen Begründung.* Frankfurt: Suhrkamp, 1990 [1918].

CHAPTER SEVEN

# Honours Without Impact: Emil von Behring's Inconsequential Nobel Prize

*Christiaan Engberts*

**Abstract**

Emil von Behring was the first recipient of the Nobel Prize in Physiology or Medicine. This had surprisingly little impact on his career or on how he was remembered after his death. This is best explained as the result of Behring's failure to stage the event. Because he had developed a reputation for being unlikeable and domineering among his colleagues, they were reluctant to celebrate his Nobel Prize with him. At the same time he had gained a lofty reputation as the 'savior of the children' among the general public, and the Nobel Prize could hardly add to this exalted public reputation.

**Keywords:** Emil von Behring, physiology, medicine, reputation

Today most people, inside as well as outside academia, agree there might not be any greater scientific honor than receiving a Nobel Prize. It has become almost unthinkable that such a professional triumph would be ignored or downplayed in any reflection on the life stories of its proud recipients. In this light, it is remarkable that most early biographers of Emil von Behring, the first recipient of the Nobel Prize in Physiology or Medicine, pay only little or even no attention to this supposed highlight of Behring's career. Paul de Kruif's popularizing account of Behring's greatest accomplishments, published almost a decade after the latter's death, passionately stresses his brilliance without even once mentioning that widely recognized sign of genius that is the Nobel Prize.[1] A detailed biography, that Heinz Zeiss and Richard Bieling published fourteen years later, includes some observations about his acceptance speech but no reflection on the impact of the Nobel Prize on (the perception of) Behring's career.[2] In George Nuttall's 1924 biographical sketch, the Nobel Prize is mentioned as an afterthought in a short paragraph listing some of the scientific and state honours that Behring received throughout his life.[3] A three page German celebration of his sixtieth birthday in 1914 – when Behring was still alive and well – does not mention his Nobel Prize at all![4]

The lack of interest among Behring's early biographers in this moment of professional triumph invites the question of why his Nobel Prize success seems to have been perceived as no more than a minor footnote to an otherwise successful and eventful scientific career. The argument that the earliest Nobel Prizes did not yet have the standing that they would develop in later years might seem plausible at first. One recent study notes that international press coverage of the first ceremony was indeed limited: specialized journals mentioned such awards in their notices, but did not dedicate full articles to them.[5] However, this argument is not entirely convincing. In the first place, the biographers mentioned in the first paragraph all wrote their assessments at a time when the prestige of the Nobel Prize had already been firmly established for many years. Secondly, the desirability of the Nobel Prize had been very clear from the beginning onwards. Contemporary newspaper even used terms like "the Olympics of science" in their early coverage.[6] The large sum of money that was awarded to the Nobel Prize's recipients also illustrated its standing. The check that Behring sent to his mother-in-law in Berlin in 1901 had a value of no less than 169,513 Mark.[7]

In the following sections, I will draw on the suggestion of Nils Hansson, Thorsten Halling, and Heiner Fangerau to look at the effect of a prize as the result of the way in which it is "staged, performed, and celebrated" to understand the lukewarm reception of Behring's Nobel Prize.[8] The authors distinguish a "behind the scenes" preceding the prize ceremony, the ceremony or act "on stage" itself, and an "after show party," when the meaning and significance of the event are further shaped by the laureates, the media, and the public.[9] I will argue that courses of events and personal relations pertaining to the "behind the scenes" and, especially, to the "after show party" of Behring's Nobel Prize success contributed to the relative neglect of his triumph during – and immediately after – his lifetime. Before I delve into the different elements of (not) staging and performing his Nobel Prize, however, I will first provide some short biographical note of Behring's life. Next, I will pay special attention to the way in which he used foreign recognition to further his career in Germany in the years before the beginning of the twentieth century. This section will emphasize especially the vigour with which he made use of foreign acclaim in the 1890s. In the light of this vigour, the significance of the question why his Nobel Prize success remained such a minor chapter in his biography during his lifetime will become increasingly clear.

In the subsequent sections, I will discuss the contexts within which Behring's career developed and the way in which they shaped the place that his Nobel Prize would assume in his reputation in his home country. After discussing concisely Behring's selection for the Nobel Prize and the subsequent ceremony in Stockholm, I will examine more closely Behring's – often strained – relations with his German scientific peers. Next, I will investigate Behring's reputation as a heroic and successful fighter of deadly disease among a broader public. The combination of the weakness of his ties to his medical and bacteriological peers, the strength of his ties to the Prussian state, and the allure of his heroic reputation among a broader audience will shed a light on the question of why neither Behring nor others were particularly keen on drawing on his Nobel Prize as a primary reason for recognizing his excellence. His excellence was, after all, widely recognized. The career overview celebrating his sixtieth birthday cited above was not the only publication to praise his "creative genius" and the "grand reshaping" of medical knowledge and practice that was his most famous accomplishment, the invention of serum therapy.[10] Finally, I will briefly reflect on how the way in which Behring was able to stage his Nobel Prize success sheds light on ideals of good scholarship and the appreciation of specific scholarly personae in early-twentieth-century Germany.

## Early success and international acclaim

Emil Behring was born in 1854 in Hansdorf, a village in the Province of Prussia in modern-day Poland.[11] He studied medicine in Berlin, where he earned his doctorate in 1879, and his licence to practise medicine in 1880. Subsequently, he found employment as an army doctor in eastern Prussia. In this capacity he was stationed at the Pharmacological Institute in Bonn between 1887 and 1889, after which he was dispatched to Robert Koch's (1843-1910) Hygienic Institute at the University of Berlin. Because he proved to be a valuable collaborator on the institute's research programs, he followed Koch when he was appointed as director of the newly established Institute for Infectious Diseases in 1891. It was at this institution that Behring carried out most of the research that would eventually result in his Nobel Prize success ten years later. As one of Koch's most trusted and qualified assistants, he had the opportunity to do extensive and innovative research into the development of a blood serum against diphtheria.

In this paper, I will not delve into the methodological and technical details of Behring's work.[12] Since the assessment of a researcher's excellence and the reception of a Nobel Prize take shape – at least to some extent –through interactions with and between their peers, I will study more closely some of the colleagues who contributed to Behring's success. Even if Koch was his supervisor during his years in Berlin, he was not his closest collaborator on the diphtheria blood serum. At least three people at Koch's research institutes stand out as key contributors to Behring's research programme: Shibasaburo Kitasato (1853-1931), Erich Wernicke (1859-1928), and Paul Ehrlich (1854-1915). Kitasato collaborated with Behring on the development of blood serums at Berlin's Hygienic Institute. The parallel work of Behring and Kitasato on diphtheria and tetanus would lay the groundwork of their new approach towards immunity that was first made public in a co-authored article in 1890.[13] Wernicke was Behring's closest collaborator after he had followed Koch to the Institute for Infectious Diseases and Kitasato had returned to his home country of Japan.[14] Wernicke's contributions were crucial: while Behring struggled with poor health, Wernicke attended to most of the all-important animal testing.[15] Finally, the contributions of Paul Ehrlich, another researcher at the Institute for Infectious Diseases in the early 1890s, made it possible to measure the efficacy of the serum accurately, and to establish the correct dosing for the safe and effective treatment of humans.[16]

The culmination of Behring's investigations came in 1894 at a meeting at the Imperial Health Office, attended by fifteen doctors, the office's director Karl Köhler (1847-1912), and Friedrich Althoff (1839-1908), the official in charge of university affairs at the Prussian Ministry of Education. It was decided that the diphtheria serum could be made commercially available at pharmacies across the country.[17] Since Behring had already established a relationship with August Laubenheimer, a member of the board of directors of the *Höchster Farbwerke*, the serum could be made available before the end of the year: the festive opening of the production facility was on 24 November 1894.[18] At this point, Behring had arguably already reached the high point of his career, but this was not reflected in his continued junior position at the Institute for Infectious Diseases. In his correspondence with Friedrich Althoff, he was very clear about the unsustainability of this arrangement, as well as of his own ambitions. In February 1895 he underlined that it was no longer possible "to continue his association with Koch's institute in the old way" and emphasized that a professorship at a Prussian university would be his preferred solution to the increasingly tense relationship with his supervisor.[19]

Behring's demand put Althof in a difficult position, because there was only a very limited number of German professorial chairs that suited Behring's research interests. Behring was also held back in his search for a professorship by his limited teaching experience. So even if Althoff was obviously sympathetic to his cause, he was not able to satisfy Behring's persistent requests as quickly as the latter had been hoping for. At this moment Behring realized the value of the recognition of his peers abroad. In almost all his letters to Althoff he drew attention to the acclaim – and even job offers – he received from representatives of foreign universities and research institutes. His relations with the *Institut Pasteur* were particularly cordial. The Parisian institute's Émile Roux (1853-1933) had been the first to isolate the diphtheria toxin and had played a major role in the clinical implementation of the serum.[20] Behring was also friendly with Roux' colleague, and future Nobel Prize recipient, Élie Metchnikoff (1845-1916). His first urgent appeal to Althoff to find a Prussian professorial chair for him, was ostentatiously written from Paris.

He underlined his warm reception at the *Institut Pasteur*: "I cannot deny that the warm recognition, that I have found here so far, has done me a lot of good; and I have promised to prolong my [...] stay in Paris in the light of the planned expressions of support."[21] Almost two weeks later he reported that his "annoyance with the lack of [German] recognition" was disappearing in the light of his warm reception by representatives of the *Institut Pasteur*, the *Académie Française*, and even the French government.[22] The government had been involved because Roux had insisted that if he were to be made an *Officier de l'Ordre National de la Légion de Honneur* for his work on the diphtheria serum, Behring should certainly receive this same distinction![23]

Behring not only used foreign praise to emphasize that he was worthy of a full professorship at a Prussian university, but he also used job offers of foreign universities and research institutes to put pressure on Althoff. If the Ministry of Education could not provide him with his desired professorial chair, he would simply accept a position elsewhere! In his letter from Paris, he mentioned an offer from the Hungarian state to establish and supervise an institute for serum production. If he was not appointed appointed to a Prussian professorship, Behring added, "I would find myself in the predicament to look for a professorship abroad and to resort to Hungary for now."[24] He soon added that he had also been invited for a similar position in Russia: "At a dinner, that I attended with the family of prince Oldenburg in Nice, I got the impression, that I can find what I need in Russia."[25] A few weeks later he added even more pressure by stressing that he could find his closest collaborators great jobs "[i]n Paris, in Rome, in Genoa, in Petersburg, in London, in Budapest, [and] in America."[26]

Behring's continuous emphasis on foreign recognition and job offers eventually paid off. In April 1895, Althoff forced the unwilling Faculty of Medicine of the University of Marburg to appoint Behring.[27] In the light of this success, by the end of the late nineteenth century, Behring was aware of the ways in which foreign recognition could help him to find recognition at home as well, and – at least as important – to achieve his professional goals. As a full professor in Marburg, he would continue to engage in new research, with a particularly strong emphasis on contributing to a field in which his former supervisor and current rival Robert Koch had failed earlier; the treatment of tuberculosis.[28] With such ambitions, all social and cultural capital that could be converted into financial support could be very useful. Therefore, it is striking that Behring scarcely tried to capitalize on his Nobel Prize victory. In the following sections I will elucidate this surprising fact in the light of Behring's professional isolation in Germany, on the one hand, and his pre-existing heroic image among the broader public, on the other.

## Receiving the Nobel Prize

Behring's reception of the Nobel Prize clearly illustrated his standing outside of Germany. Not only was the prize itself an honour from abroad, but all his nominations were by foreign scholars as well. This was remarkable, because other scholars with many nominations tended to be championed by their compatriots. The Russian physiologist Ivan Pavlov (1849-1936), who had received the most nominations, was primarily supported by members of Petersburg's Faculty of Military Medicine, while the runner-up, the Spanish neuroscientist Santiago Ramón y Cajal (1852-1934) relied heavily on the votes of his colleagues from Madrid.[29] Behring, however, was only nominated by his peers from Switzerland (Berne), Norway (Kristiana), Hungary (Budapest), and the Netherlands (Leiden). None of the people who suggested Behring seem to have been personally close to him. The Swiss anatomist Theodor Langhans (1839-1915) had worked in Marburg for some years but had already been teaching in Berne for more than two decades when Behring arrived there. The Norwegian and Hungarian nominators may have been oriented towards German scholarship in general, but had no personal relationship with Behring, either. The same is true for the nine Leiden professors, who collectively nominated Behring for the Nobel Prize.[30] Surprisingly, none of his admiring friends at the *Institut Pasteur* nominated Behring.[31]

The Nobel Committee only agreed on Behring after long deliberations. The struggle to reach an agreement is best understood as the result of the criteria set out in Alfred Nobel's will: the prize should be awarded to honour discoveries that were first recent and groundbreaking, and second of the "greatest benefit to mankind."[32] The Committee was initially divided among advocates of Ronald Ross's (1857-1932) groundbreaking discovery of the life cycle of the causative agent of malaria, and proponents of Niels Finsen's (1860-1904) new phototherapy for skin diseases. The former was the more groundbreaking theoretical innovation, while the latter was more easily applied in practice and could therefore be seen as an invention of greater immediate benefit to many people.[33] Behring's blood serum therapy provided the perfect compromise. After all, it was both a major theoretical innovation in the study of immunity and obviously of great benefit to those who suffered from this deadly disease. Count Karl Mörner (1854-1917) emphasized both virtues in his speech at the ceremony: "Up until now, serum therapy has had particularly splendid triumphs in the case of diphtheria, but its significance is not limited to this disease [...]. The field which is opened up for research by the development of serum therapy has [...] no discernible limits."[34]

Even if nobody at the *Institut Pasteur* had nominated Behring, the warmest congratulations immediately came to him from Paris. While Behring was still enjoying the Stockholm celebrations – and before he had even given his acceptance speech

– both Metchnikoff and Roux wrote to congratulate their Marburg colleague. "I hurry to express to you my most heartfelt and sincere congratulations for this high distinction, which you have deserved so fully," wrote Metchnikoff.[35] Émile Roux was even more jubilant on behalf of his whole institute: "We are all cheering at the *Institut Pasteur* at the decision of the Nobel Committee, which rewards your marvelous [sic] discoveries and the great service that you have rendered to mankind."[36] His German peers seem to have been less keen to congratulate their compatriot, even if he had earlier received a short but enthusiastic telegram from Althoff, who sent him his greetings "with a joyous hurrah."[37]

Though the ceremony and the congratulations from Paris reaffirmed Behring's international standing, his acceptance speech showed subtle signs of his less cordial relations with his German peers. After a reflection on the success of his diphtheria serum in the first half of his speech, Behring discussed his more recent research into bovine tuberculosis in its second half. He used the festive occasion of the award show to distance himself from the findings of his former supervisor Koch. The latter had suggested that tuberculosis bacteria might not be the same in humans and cattle. Behring framed most of his Stockholm remarks as a presentation of the results of experiments on goats and cattle, that proved that recently cultivated human tuberculosis cultures were "by no means unharmful" to cattle.[38] To German insiders it was clear that Behring's decision to single out Koch for criticism at this occasion was not part of a friendly, scientific debate. This was especially clear in the light of the fact that the break between them had been solidified by a long litigation process at the Patent Office, in which Koch had unsuccessfully challenged the patentability of a tuberculosis antitoxin developed in Behring's Marburg laboratory.[39]

### Isolation among his peers

To many of his German peers, Behring's use of the festive ceremony to engage in such a loaded debate with a highly respected colleague, must have appeared as a typical example of his bellicose character. This aspect of his personality had earlier complicated Althoff's efforts to find a university willing to appoint him as a full professor. When in 1894 a Berlin newspaper reported early successful experiments with the diphtheria serum by the Berlin physician Rudolf Virchow (1821-1902), Behring felt his priority claim was threatened and reacted furiously. Since Virchow had earlier been critical of Behring's theoretical underpinning of the serum's efficacy, Behring chose to frame him as a "medical doctrinaire," whose ideas about the causes of diseases were "heresies," and whose "dogmatism" had resulted in an "inquisition." His serum could certainly never have been developed under the supervision of Virchow![40] Meanwhile, Virchow denied the newspaper's reports and recommended further research into Behring's serum, that he had happily and successfully administered at Berlin's *Kaiser- und Kaiserin-Friedrich-Kinderkrankhaus*.[41] Althoff's informant in Marburg, Carl Fraenkel (1861-1915), told him that Behring's unnecessarily hostile demeanour towards Virchow had been one of the main reasons why his peers at the Faculty of Medicine preferred to appoint another colleague.[42]

As the soured relationship with his former supervisor Koch suggests, Behring's interaction with his close collaborators could also be awkward. The deterioration of his relationship with Koch was certainly not only Behring's fault. After all, Koch's impulse to protect his own primacy in the fight against tuberculosis by challenging Behring's patents was a major contributing factor to the disintegration of their cordial relationship. However, Behring's troubles with maintaining amicable ties with his colleagues could not always be blamed on the others. This can be illustrated by a closer look at his relationships with his peers who had made major contributions to the development of the diphtheria serum. The only relationship that did not sour was the one with Kitasato. Kitasato had returned to Japan in the early 1890s, and he soon established his own Institute for Study of Infectious Diseases there. At his departure, he let Behring know that he counted "the time of my collaborative work with you among the most enjoyable hours of my stay in Berlin."[43] A few years later he would send one of his most promising students, Taichi Kitashima (1870-1956), to Marburg to study with Behring.[44] Kitashima would stay in Marburg for four years, seemingly to the satisfaction of both Behring and Kitasato. He would later succeed the latter as director of the Tokyo research institute, and would stay on friendly terms with his German teacher. Ten years after Behring's death Kitashima would still visit Behring's widow on a trip to Europe.[45]

Behring's relationships with his collaborators who stayed in Germany, Erich Wernicke and Paul Ehrlich, were much more complicated. Wernicke was five years younger than Behring and when he first arrived at the Institute for Infectious Diseases he was "caught by Behring's towering idiosyncratic character."[46] In the early 1890s, their relationship was very intimate. Behring even suggested to share an apartment and observed that they were "already semi-married, after all."[47] It would not be long, however, for their relationship to take a turn for the worse. Wernicke soon started to accuse Behring of bossing him around, to which Behring curtly reacted with the admonition to please write him "many, but less reproachful, letters."[48] A few years later, their relationship would sour even further, when Wernicke complained about the meagre financial rewards he received for his indispensable contributions to the commercially very successful diphtheria serum. Behring rebuked him by simply claiming to own the full "scientific and financial rights of discovery."[49] Their relationship only improved when Wernicke later moved far away from Marburg after he had been appointed as director of the Hygienic Institute of Posen in 1899.

Like Wernicke, Ehrlich was disappointed about his rewards for his contributions to the diphtheria serum. Behring had convinced him to abstain from a share of the serum's profits, and had promised that he would arrange a secure, permanent, position as director of an independent serum testing institute for him.[50] The establishment of this institute and the appointment of Ehrlich took longer than he had hoped for, and, once he had started in this new position, he was dissatisfied by the fact that Behring seemed to conceive of it as a state-sponsored support organization for his own research programme rather than as a truly independent testing institute. The relationship between Behring and Ehrlich never fully recovered. Even Althoff could not fix the unease between the two great bacteriologists. When Althoff pressed Ehrlich to work with Behring again, Ehrlich curtly replied that collaborating with Behring could only be expected from "a slavish character, but not from an independent researcher with the greatest thirst for freedom (such as I am, after all)."[51] Through the years, Behring would lose more professional friends. Eventually, he would even admit to Wernicke that "among the many reproaches I make myself, one of the most severe is that through the years I have lost one friend after the other by my own fault."[52]

## Scientific hero or benefactor of humanity?

Behring's troubled relations with his German peers limited the ways in which he could stage his Nobel Prize reception. His colleagues were largely unwilling to recognize him proudly as a shining example of the internationally recognized greatness of German science. Most of the few compatriots who were willing to celebrate his Nobel Prize with him were not prominent scientists themselves. Althoff, even if he maintained close relations with several prominent medical scholars, had no training in medicine. Julius Holtz (1836-1911), who sent his congratulations to Else Behring (1876-1936) while her husband was in Stockholm, was a representative of the pharmaceutical industry rather than of the academic medical community.[53] Ernst Schweninger (1850-1924) was one of the few medical professors to express his happiness about Behring's Nobel Prize.[54] However, Schweninger's reputation among his peers was even worse than Behring's: while they tended to dislike Behring's character but did not doubt his capabilities, Schweninger was seen as both an unpleasant person and an unqualified scholar.[55] Without the support of his scientific peers, it was neither easy nor obvious for Behring to present himself as a proud paragon of German science. And since the Nobel Prize in Medicine had also been awarded based on its recipient being of the "greatest benefit to mankind," it made more sense for Behring to present himself as a great benefactor of humanity rather than as a great German scientist.

After all, in the eyes of a broader, non-academic, audience, Behring's Nobel Prize did not add significantly to his already stellar reputation. Even if few members of this wider audience may have cared about this twentieth-century investigations into bovine tuberculosis, he was loved and admired for the success of his diphtheria blood serum. Throughout the years, Behring received thankful letters from parents whose children had been cured from diphtheria by his invention. In the light of the high mortality rate among children with diphtheria before the invention of the serum, the tremendous thankfulness of the families was understandable. A quick look into the discourse of gratitude in the letters that Behring continued to receive throughout the late 1890s and the early 1900s, illustrates how little the Nobel Prize could add to the lofty popular image of Behring as "the benefactor of humanity" and "the saviour of the children."

A letter of thanks Behring received from a grateful Moscow father in November 1901 is quite representative of the kind of praise he received: "I had the opportunity to observe how the serum – an everlasting monument for your honour for eternity – works like a miracle, and I believe I am not alone when I allow myself to put my gratitude for the saviour of countless lives, the comforter of innumerable, saddened hearts of parents […] into words."[56] This emphasis on the extent to which Behring's serum had benefited not just the families of the individual letter writers, but humanity at large, was a recurring theme. Another Moscow father was even more jubilant than the one cited above: "My soul is so full of the most profound gratitude for you, whose great discovery has given the wretched people such a remarkable remedy to fight diphtheria, this dreadful scourge of humanity. […] The thought about how much you have done for wretched humanity, as well as boundless gratitude make me cry tears of the most profound emotion."[57]

One way for grateful parents to express the profoundness of their thankfulness was by sending unsolicited presents to Marburg. The first Moscow father quoted above added 100 Mark to his letter. He left Behring free to donate this money to a charity of his own choice. One week before Christmas 1899, the grateful painter Leo Reiffenstein (1856-1924) sent Behring a painting that he had named *Weihnachtslied* (Christmas Song): "I do not, alas, know whether it suits your taste and whether it will be able to give you the joy that we hope for; may you just recognize the desire to give you a minor favour in return […] for the great blessing that you have given all of humanity."[58] Another way in which parents expressed the depth of their admiration for Behring was by using language that bestowed an almost godly status on him. A professor at the Prussian military academy wrote: "While he was growing up to be a promising young man, my child has always been reminded that, next to God, he should always gratefully recognize you as the provider of his new life."[59]

A special category among the many letters of thanks that Behring received were those written by recovered children themselves. Sometimes these letters drew on the discourse of Behring's God-like status: "I am 10 years old; had a terrible diphtheria and would surely have died, if the doctor would not have injected me with your serum. I owe my life only to God and to you."[60] A few months before Behring heard about his Nobel Prize, a German-American girl was less lofty but not less grateful in the note she sent: "I am eight years old. In March I was very sick and now I am healthy and happy again; my dad told me that your medicine has cured me and therefore I would like to thank you. [...] I am tired. Good night, dear *Herr Doktor*, I would love to know what you look like."[61] Another young girl wrote Behring in a somewhat more formal style but was as thankful as the others: "Forgive me if I, as a young girl unknown to you, thank you sincerely for inventing the serum. I was seriously ill with diphtheria this year and I definitely owe my life to the immediate use of the serum! I could not resist the wish to send you my most sincere gratitude."[62]

The praise for Behring in the popular press consisted of discourses of Behring as benefactor of humanity and saviour of the children rather than those of scholarly excellence we tend to associate with the Nobel Prize for Medicine in the twenty-first century. Even one of his closest scientific collaborators in Marburg, Paul Römer (1876-1916), emphasized how Behring's work has been of the greatest benefit to mankind over a discussion of his scientific genius by titling a celebratory review of Bering's career "A Benefactor of Humanity."[63] Römer's review was written in honour of Behring's sixtieth birthday. The local newspaper also reported on the festivities and underscored that Behring's work not only had scholarly merit but also had great "merit for humanity."[64] The newspaper's reporting did not go into any detail about Behring's scholarly accomplishments, but focused mostly on the celebrations itself, highlighting the floral arrangements, the congratulations from local gymnastic societies (*Turnvereine*), Behring's honorary citizenship of the city of Marburg, and the congratulatory address presented by the dean of Marburg's Faculty of Medicine. The newspaper further emphasized that the certificate for Behring's honorary citizenship, the imagery added to the Faculty's address, and a celebratory postcard had all been designed by local artist Otto Ubbelohde (1867-1922). All things considered, the local reporters wrote about Behring as a local celebrity rather than as a medical researcher of international renown.

### (Not) Staging a Nobel Prize

Two of the more complex and compelling reasons why Behring's Nobel Prize victory was scarcely used as an opportunity to stage an image of scientific excellence in his home country have been treated in-depth above. First, Behring was a bit of an outcast in the circles of his medical and bacteriological peers. Second, his popular image as a noble benefactor of mankind was already well-established by the time he received his Nobel Prize, which lowered the need to lean into the image of scientific excellence that his Nobel Prize success would have allowed him to stage. While discussing these two points, I have also touched on some other circumstances that may have shaped Behring's ability and willingness to use his Nobel Prize victory as an opportunity to underline his scientific excellence in the eyes of both his peers and a broader audience. For a more comprehensive view of the contexts that contributed to the insignificance of the Nobel Prize in early accounts of Behring's life, account must be taken of two circumstances that I have only hinted at above.

The first of these circumstances was suggested in the discussion of Behring's reliance on foreign acclaim to force Althoff at the Ministry of Education to secure him a full professorship at a Prussian university. Behring's willingness to pressure Althoff was the result of two things. First, he was conducting his research at Koch's Institute for Infectious Diseases, which was bound to eventually limit his freedom to engage in his own research. Second, he was not yet sure that Althoff would be willing and able to find him the professorial chair he desired. When he received his Nobel Prize almost a decade later, all these limiting circumstances had been handled. Since he was secure in his position as a full professor at Marburg, he no longer had to worry about his research priorities being steered by anyone else. In addition, Althoff had proven to be a trustworthy and capable ally at the ministry, which took away the perceived need to use foreign praise to put pressure on him. Althoff even became a friend of the Behring family. In 1903 he even accepted the Behrings' invitation to be godfather to their third son, Hans Adolf.[65]

The second way in which Behring's life was very different in 1901 than it had been in the early 1890s was also suggested above, when I drew attention to his collaboration with the *Höchster Farbwerke* to make his diphtheria serum available at pharmacies across the nation. The mass production of his serum not only impacted his life by provoking awkward disagreements about financial compensation with his close collaborators Wernicke and Ehrlich, it also made him a rich man with warm relationships with the pharmaceutical industry. During the first year of its production, the serum generated a profit of almost 450,000 Marks, and in the second year this number rose even further to more than 750,000 Marks.[66] Behring's contract at the time stipulated that he would receive half of these profits, so between 1894 and 1896 alone his serum made him more than half a million Marks.[67] Because of this commercial success, Behring's relationships with the industry increased while his relationships with his academic peers dramatically deteriorated. The new financial security also ensured that he would be much less dependent on other parties' willingness to invest in his endeavors than ever before, which reduced the need to put additional effort into using his Nobel Prize to stage himself as an excellent, groundbreaking researcher, who both needed and deserved financial support to carry out his new promising research projects.

With these final additions, it has become clear why receiving the Nobel Prize seems to have had so little impact on the life and early perception of Emil Behring.

There was no evident, straightforward way to use his victory to stage himself as an exemplar of scientific excellence. This was the case both because there was only limited *opportunity* to do this, and because there was very little *need* to do this. The opportunity was limited because Behring had set his mind on having a career in Germany. However, his relationships with most other prominent medical and bacteriological researchers in Germany were badly tarnished. Therefore, the audience most likely both to recognize the groundbreaking merit of his work and to validate his scientific excellence in the eyes of a broader audience, was reluctant to celebrate his Nobel Prize with him. In the years after his Nobel Prize, Behring did not notice an increase in his colleagues' willingness to engage with him. On the contrary, he soon started noticing that he was occasionally no longer even invited to scientific conferences in his own field of research![68] By 1906, Behring even characterized himself as an "outsider" in "our medical university life" in a disheartened letter to Althoff.[69]

The lack of possibilities to stage his excellence in front of an audience of his peers, was to some extent compensated by a number of circumstances that together removed the need to seek their recognition. First, since the 1890s he had developed a strong reputation not primarily as an excellent researcher but as a civic hero: a benefactor of mankind and a saviour of the children. Among a broad public, he enjoyed a reputation that was even loftier than that of just another excellent German academic. Second, he no longer needed a reputation for scientific excellence to continue his research on his own terms. He had the secure basis of a permanent position as a full professor, an extraordinarily good relationship with the most influential individual in the field of university policy at the Ministry of Education, and a good working relationship with the pharmaceutical industry. He could not expect to gain a great deal by staging himself as an excellent researcher on the basis of his Nobel Prize success: apart from the sympathy of his peers, he had everything he could wish for.

Behring's modest efforts and success in staging his election for the Nobel Prize highlight both his outsider status among his peers and his heroic image among a broader audience. This also sheds light on early-twentieth-century ideals of good scholarship. The concept of the scholarly persona is particularly suited to reflect on the relation between individual biography and shared conceptions of what it means to be an exemplary scholar. This becomes particularly apparent in the light of the fact that this concept allows both for the investigation of conceptions of scholarly virtue as they are shared in scholarly communities and for the search for the ways in which individual scholars stage their careers and fashion themselves as deserving members of such communities.[70] The lukewarm reception of Behring's Nobel Prize draws attention to the existence of shared ideals of scholarship, as well as the ways in which such ideals are intertwined with the individual scholar's ability and willingness to perform them.

In breaking – or at least failing to live up to – some of the norms of good scholarship, Behring's example illustrates the scholarly virtues that were most highly valued in early-twentieth-century German academia. To be sure, he did live up to some weighty expectations: nobody denied, for example, the originality of his thought or his work ethic. He fell short, however, in regard to the norms and virtues that shape the social texture of the scholarly world. His harsh public attacks of Virchow suggested a lack of both respect and modesty; his relationships with (former) collaborators could be interpreted as lacking willingness to acknowledge others' contributions to his success, as well as a lack of personal loyalty; his attempts to control an independent serum testing institute were seen by its director Ehrlich as a threat to his intellectual independence.[71] It is worth noting, however, that none of this damaged his reputation among a broader audience. Apparently, public and academic ideals of good scholarship did not overlap entirely. In light of the popular admiration for his accomplishments, Behring was able to ignore academic conceptions of scholarly virtue at least to some extent. However, by refusing to embody the persona that his academic peers expected him to adhere to, by embracing his reputation as a civic hero at least as much as a scientific hero, he limited his opportunities to stage his Nobel Prize as a scientific triumph.

Therefore, the decision of the Nobel Committee to give its first Nobel Prize in Physiology or Medicine to Behring, hardly had any impact on the latter's career. The committee's selection seems to have been fortunate, however, in at least one way. The bar that had been set for this prize was high: its recipient was expected not only to have made a discovery that was both recent and groundbreaking, this discovery was also expected to have been of the "greatest benefit to mankind." Only very few groundbreaking discoveries prove to be of the greatest benefit to mankind while still being recent. With his recent, groundbreaking, child-saving diphtheria serum, Behring gave the Nobel Committee the prize recipient it was looking for. In retrospect, Behring may have done more for the prestige of the Nobel Prize than the Nobel Prize has done to (the memory of) his career.

## Primary sources

Behring-Nachlass digital

Accessible online at: https://evb.online.uni-marburg.de/cgi-bin/evb?t_systematik=x

Geheimes Staatsarchiv Preußischer Kulturbesitz

VI. HA, Nl Althoff, F. T., No. 325, v. Behring. Heilseruminstitut. Briefe v. Behrings

VI. HA, Nl Althoff, F. T., No. 326, v. Behring. Heilseruminstitut. Allgemeine Briefe

VI. HA, Nl Althoff, F. T., No. 668, Behring-Bellermann

VI. HA, Nl Althoff, F. T., No. 709, Ehrenberg-Ehrlich

## Notes

1 Paul de Kruif, *Microbe Hunters* (New York: Blue Ribbon Books, 1926), 184–206.

2 Heinz Zeiss and Richard Bieling, *Behring: Gestalt und Werk* (Berlin-Grunewald: Bruno Schultz, 1940), 276.

3 George H.F. Nuttall, "Biographical Notes Bearing on Koch, Ehrlich, Behring and Loeffler, With Their Portraits and Letters from Three of Them," *Parasitology* 16 (1924), 232.

4 Hans Much, "E. von Behring. Ein Wort zu seinem sechzigsten Geburtstage," *Berliner klinische Wochenschrift* 11 (1914), 1–3.

5 Derek S. Linton, *Emil von Behring: Infectious Disease, Immunology, Serum Therapy* (Philadelphia: American Philosophical Society, 2005), 238.

6 Nils Hansson and Ulrike Enke, "Emil von Behring: erster Nobelpreisträger für Medizin – Die Bekämpfung der Diphterie," *Deutsche medizinische Wochenschrift* 140 (2015), 1898. The comparison between the Olympics and the Nobel Prize is particularly convincing because the first modern Olympics took place only five years before the first Nobel Prize ceremony. In both cases its international appeal was clear from the beginning.

7 Emil von Behring to Elise Spinola, 12 December 1901, Behring-Nachlass digital (hereafter BNd): EvB/B 214/32.

8 Nils Hansson, Thorsten Halling and Heiner Fangerau, "Introduction," in *Attributing Excellence in Medicine: The History of the Nobel Prize*, eds. Nils Hansson, Thorsten Halling and Heiner Fangerau (Leiden: Brill, 2019), 5.

9 Ibid., 3.

10 Much, "E. von Behring," 2.

11 These biographical details are based on Linton, *Emil von Behring*, 17–40 and Zeiss and Bieling, *Behring*, 13–53.

12 For a concise overview of the way in which Behring's serum built and innovated on the research of his peers, see: Christoph Gradmann, "Locating Therapeutic Vaccines in Nineteenth-Century History," *Science in Context* 21 (2008), 145–160.

13 Emil Behring and Shibasaburo Kitasato, "Über das Zustandekommen der Diphtherie-Immunität und der Tetanus-Immunität bei Thieren," *Deutsche medizinische Wochenschrift* 16 (1890), 1113–1114. Emil Behring was elevated to nobility in January 1901, which allowed him to call himself 'von Behring. I will refer to 'Emil Behring' when citing sources from before January 1901 and to 'Emil von Behring' when citing later sources. Some more details about the collaboration between Behring and Kitasato are provided in Franz Luttenberger, "Excellence and Chance: The Nobel Prize Case of E. von Behring and É. Roux," *History and Philosophy of the Life Sciences* 18, (1996), 225–227.

14 An extensive investigation of the collaboration between Behring and Wernicke can be found in Erika Schulte, *Der Anteil Erich Wernickes an der Entwicklung des Diphtherieantitoxins* (Berlin: Mensch & Buch, 2001).

15 Erich Wernicke to Bernhard Möllers, 29 August 1924, NBd: EvB/F 5.

16 Axel C. Hüntelmann, *Paul Ehrlich: Leben, Forschung, Ökonomien, Netzwerke* (Göttingen: Wallstein Verlag, 2011), 91–108; Ernst Baumler, *Paul Ehrlich: Forscher für das Leben*, 3. Durchgesehene Auflage (Frankfurt am Main: Edition Wötzel, 1997), 92–93.

17 Beratung betreffend das Diphtherieserum [o. Datum; sicher aber am 3.1.1895], BNd: EvB/B196/5.

18 Linton, *Emil von Behring*, 179.

19 Emil Behring to Friedrich Althoff, 8 February 1895, Geheimes Staatsarchiv Preußischer Kulturbesitz (hereafter GStA PK), VI. HA, Nl Althoff, No. 325.

20 Luttenberger, "Excellence and Chance," 238.

21 Emil Behring to Friedrich Althoff, 24 December 1894, GStA PK, VI. HA, Nl Althoff, No. 325.

22 Emil Behring to Friedrich Althoff, 2 January 1895, GStA PK, VI. HA, Nl Althoff, No. 325.

23 Ulf Lagerkvist, *Pioneers of Microbiology and the Nobel Prize* (New Jersey: World Scientific, 2003), 99.

24 Emil Behring to Friedrich Althoff, 24 December 1894, GStA PK, VI. HA, Nl Althoff, No. 325.

25 Emil Behring to Friedrich Althoff, 24 January 1895, GStA PK, VI. HA, Nl Althoff, No. 325.

26 Emil Behring to Friedrich Althoff, 8 February 1895, GStA PK, VI. HA, Nl Althoff, No. 325.

27 I have described this process in more detail in Christiaan Engberts, *Scholarly Virtues in Nineteenth-Century Sciences and Humanities: Loyalty and Independence Entangled* (Cham: Palgrave MacMillan, 2022), 153–156.

28 For Koch's failed attempts to find a cure for tuberculosis, see: Christoph Gradmann, "Robert Koch und das Tuberkulin – Anatomie eines Fehlschlags," *Deutsche medizinische Wochenschrift* 124 (1999), 1253–1256. For more on Behring's tuberculosis research, see: Linton, *Emil von Behring*, chapter 6.

29 Ulrike Enke, "„Der erste zu sein.“ – Über den ersten Medizinnobelpreis für Emil von Behring im Jahr 1901," *Berichte zur Wissenschaftsgeschichte* 41, No. 1 (2018), 24. Even though Enke identifies Ramón y Cajal as receiving the second-most number of nominations, Luttenberger identifies Behring as the runner-up and suggests that Ramón y Cajal received fewer votes than Ronald Ross, who received seven: Luttenberger, "Excellence and Chance," 229.

30 Enke, "Der erste zu sein," 24–25.

31 Nor did anyone at the *Institut Pasteur* nominate its very own Émile Roux, see: Luttenberger, "Excellence and Chance," 230.

32 Hansson, Halling and Fangerau, "Introduction," 1.

33 Luttenberger, "Excellence and Chance," 231–232.

34 Quoted in Scott H. Podolsky, "From Global Recognition to Global Health: Antimicrobials and the Nobel Prize, 1901-2015," in *Attributing Excellence in Medicine: The History of the Nobel Prize*, eds. Nils Hansson, Thorsten Halling and Heiner Fangerau (Leiden: Brill, 2019), 83.

35 Élie Metchnikoff to Emil von Behring, 11 December 1901, BNd: EvB/B 101/13.

36 Émile Roux to Emil von Behring, 11 December 1901, BNd: EvB/B 126/6.

37 Friedrich Althoff to Emil von Behring, 19 November 1901, BNd: EvB/B 3/17.

38 Emil von Behring, *Nobel Lecture: Serum Therapy in Therapeutics and Medical Science*, 12 December 1901, https://www.nobelprize.org/prizes/medicine/1901/behring/lecture/.

39 Linton, *Emil von Behring*, 265–268.

40 Emil Behring, "Das neue Diphtheriemittel," *Die Zukunft* 9 (1894), 97–100 and 249–264.

41 Rudolf Virchow to Friedrich Althoff, 17 October 1894, GStA PK, VI. HA, Nl. Althoff, No. 326.

42 Carl Fraenkel to Friedrich Althoff, 18 December 1894, GStA PK, VI. HA, Nl. Althoff, No. 326.

43 Shibasaburo Kitasato to Emil Behring, 27 March 1892, BNd: EvB/B 70/2.

44 Ulrike Enke and Aeka Ishihara, "Ein Japaner in Marburg: Aus den Erinnerungen – *Jiden* – des japanischen Bakteriologen Taichi Kitashima (1870-1956)," *NTM Zeitschrift für Geschichte der Wissenschaften, Technik und Medizin* 25 (2017), 239–240.

45 Private guestbook of Emil and Else von Behring, entry for 8 April 1927, BNd: EvB/L 266.

46 Erich Wernicke to Bernhard Möllers, 29 August 1924, BNd: EvB/F 5.

47 Emil Behring to Erich Wernicke, 22 November 1891, BNd: EvB/B B 1/177 and 15 February 1892, BNd, EvB/B 1/186.

48 Emil Behring to Erich Wernicke, 25 February 1897, BNd: EvB/B B 1/244.

49 Emil Behring to Erich Wernicke, 2 January 1899, BNd: EvB/B B 1/248.

50 Hüntelmann, *Paul Ehrlich*, 104.

51 Paul Ehrlich to Friedrich Althoff, 12 September 1903, GStA PK, VI. HA, Nl Althoff, No. 709.

52 Emil von Behring to Erich Wernicke, 12 December 1908, BNd: EvB/B 1/273.

53 Julius Holtz to Else Behring, 11 December 1901, BNd: EvB/B 171. For more on Julius Holtz, see his entry in *Deutsche Biographie*: https://www.deutsche-biographie.de/pnd137067216.html?language=de.

54 Ernst Schweninger to Emil von Behring, 3 February 1902, BNd: EvB/B 138/2.

55 See, for example, the remarks in Karl Ed. Rothschuh, "Ernst Schweninger (1850-1924) Zu seinem Leben und Wirken: Ergänzungen, Korrekturen," *Medizinhistorisches Journal* 19 (1984), 250–258.

56 Ludwig Mandl to Emil Behring, 15 November 1900, BNd: EvB/B 161/33.

57 N. Kovalensky to Emil von Behring, 22 December 1902, BNd: EvB/B 161/26.

58 Leo Reiffenstein to Emil Behring, 18 December 1899, BNd: EvB/B 161/37.

59 Wilhelm Julius von Goerne to Emil von Behring, 13 September 1903, BNd: EvB/B 161/15. The letter is signed "von Goerne an der Haupt-Kadetten-Anstalt." I have assumed it is Wilhelm Julius von Goerne because he is mentioned as professor at this academy at the website of the *Deutsche Gesellschaft für Ordenskunde*: Nikolai Scheuring, *Liste der Ehrenritter des Johanniterordens 1853-1918*, https://www.deutsche-gesellschaft-fuer-ordenskunde.de/DGOWP/wp-content/uploads/2021/07/Liste_Ehrenritter_Johanniterorden_N_Scheuring.pdf.

60 Ebba Othman to Emil von Behring, no date [probably around 1908], BNd: EvB/B 161/36.

61 Gerda Fischer to Emil von Behring, 18 June 1901, BNd: EvB/B 161/8. My italics.

62 Berta Schleicher to Emil von Behring, 11 September 1910, BNd: EvB/B 161/40.

63 Paul Römer, "Ein Wohltäter der Menschheit," *Reclams Universum. Moderne Illustrierte Wochenschrift* 30 (1914), 84–87. Accessed at BNd: EvB/S 5/4.

64 "Geheimrat v. Behrings 60s Geburtstag," *Oberhessische Zeitung* 49 (1914), 1. Accessed at BNd: EvB/S 5/3.

65 Else von Behring to Friedrich Althoff, 3 November 1903, GStA PK, VI. HA, Nl Althoff, No. 668; Friedrich Althoff to Emil von Behring, 14 December 1903, BNd: EvB/B 3/24.

66 Based on a note reading '*Diphterie Heilserum*': BNd: EvB/B 196/115. The profit for 1894 is also mentioned in a report by *Farbwerke* director Laubenheimer: BNd, EvB/B 196/7.

67 Vertrag zwischen Emil Behring und den Farbwerken Vormals Meister, Lucius & Brüning betreffend 'Gewinnung von Diphterie-Heilserum', § IV, EvB/B 196/2/4.

68 Ferdinand Hueppe to Emil von Behring, 13 February 1906, GStA PK, VI. HA, Nl Althoff, No. 668.

69 Emil von Behring to Friedrich Althoff, 15 February 1906, GStA PK, VI. HA, Nl Althoff, No. 668.

70 For more detail on the first approach, see: Herman Paul, "What is a Scholarly Persona? Ten Theses on Virtues, Skills, and Desires," *History and Theory* 53 (2014), 348–371. For more details on the second approach, see: Mineke Bosch, "Scholarly Personae and Twentieth-Century Historians: Explorations of a Concept," *BMGN – Low Countries Historical Review* 131 (2016), 33–54.

71 I have discussed the relationship between Behring and Ehrlich in more detail in Engberts, *Scholarly Virtues*, 193–194.

## Bibliography

Bäumler, Ernst. *Paul Ehrlich: Forscher für das Leben*, 3. Durchgesehene Auflage. Frankfurt am Main: Edition Wötzel, 1997.

Behring, Emil. "Das neue Diphtheriemittel." *Die Zukunft* 9 (1894): 97–100, 249–264.

Behring, Emil von. *Nobel Lecture: Serum Therapy in Therapeutics and Medical Science*, 12 December 1901, https://www.nobelprize.org/prizes/medicine/1901/behring/lecture/.

Behring, Emil and Shibasaburo Kitasato. "Über das Zustandekommen der Diphtherie-Immunität und der Tetanus-Immunität bei Thieren." *Deutsche medizinische Wochenschrift* 16 (1890): 1113–1114.

Bosch, Mineke. "Scholarly Personae and Twentieth-Century Historians: Explorations of a Concept, *BMGN – Low Countries Historical Review* 131:4 (2016): 33–54.

Engberts, Christiaan. *Scholarly Virtues in Nineteenth-Century Sciences and Humanities: Loyalty and Independence Entangled*. Cham: Palgrave MacMillan, 2022.

Enke, Ulrike. "„Der erste zu sein."– Über den ersten Medizinnobelpreis für Emil von Behring im Jahr 1901." *Berichte zur Wissenschaftsgeschichte* 41 (2018): 19–46.

Enke, Ulrike and Aeka Ishihara. "Ein Japaner in Marburg: Aus den Erinnerungen – *Jiden* – des japanischen Bakteriologen Taichi Kitashima (1870-1956)." *NTM Zeitschrift für Geschichte der Wissenschaften, Technik und Medizin* 25 (2017): 237–256.

Gradmann, Christoph. "Robert Koch und das Tuberkulin – Anatomie eines Fehlschlags." *Deutsche medizinische Wochenschrift* 124 (1999): 1253–1256.

Gradmann, Christoph. "Locating Therapeutic Vaccines in Nineteenth-Century History." *Science in Context* 21 (2008): 145–160.

Hansson, Nils and Ulrike Enke. "Emil von Behring: erster Nobelpreisträger für Medizin – Die Bekämpfung der Diphterie." *Deutsche medizinische Wochenschrift* 140 (2015): 1898–1902.

Hansson, Nils, Thorsten Halling and Heiner Fangerau. "Introduction," in *Attributing Excellence in Medicine: The History of the Nobel Prize*, edited by Nils Hansson, Thorsten Halling, and Heiner Fangerau, 1–14. Leiden: Brill, 2019.

Hüntelmann, Axel C. *Paul Ehrlich: Leben, Forschung, Ökonomien, Netzwerke*. Göttingen: Wallstein Verlag, 2011.

Kruif, Paul de. *Microbe Hunters*. New York: Blue Ribbon Books, 1926.

Lagerkvist, Ulf. *Pioneers of Microbiology and the Nobel Prize*. New Jersey: World Scientific, 2003.

Linton, Derek S. *Emil von Behring: Infectious Disease, Immunology, Serum Therapy*. Philadelphia: American Philosophical Society, 2005.

Luttenberger, Franz. "Excellence and Chance: The Nobel Prize Case of E. von Behring and É. Roux." *History and Philosophy of the Life Sciences* 18 (1996): 225–239.

Much, Hans. "E. von Behring. Ein Wort zu seinem sechzigsten Geburtstage." *Berliner klinische Wochenschrift* 11 (1914): 1–3.

Nuttall, George H.F. "Biographical Notes Bearing on Koch, Ehrlich, Behring and Loeffler, With Their Portraits and Letters from Three of Them." *Parasitology* 16 (1924): 214–238.

Paul, Herman. "What is a Scholarly Persona? Ten Theses on Virtues, Skills, and Desires." *History and Theory* 53 (2014): 348–371.

Podolsky, Scott H. "From Global Recognition to Global Health: Antimicrobials and the Nobel Prize, 1901-2015." In *Attributing Excellence in Medicine: The History of the Nobel Prize*, edited by Nils Hansson, Thorsten Halling and Heiner Fangerau, 81–96. Leiden: Brill, 2019.

Römer, Paul. "Ein Wohltäter der Menschheit." *Reclams Universum. Moderne Illustrierte Wochenschrift* 30 (1914): 84–87.

Rothschuh, Karl Ed. "Ernst Schweninger (1850-1924) Zu seinem Leben und Wirken: Ergänzungen, Korrekturen." *Medizinhistorisches Journal* 19 (1984): 250–258.

Schulte, Erika. *Der Anteil Erich Wernickes an der Entwicklung des Diphtherieantitoxins*. Berlin: Mensch & Buch, 2001.

Zeiss, Heinz and Richard Bieling. *Behring: Gestalt und Werk*. Berlin Grunewald: Bruno Schultz, 1940.

CHAPTER EIGHT

# A Hotly Contested Nobel: Christiaan Eijkman, Gerrit Grijns and the Discovery of Vitamin B1

*Rob van den Berg*

**Abstract**

Nobel Prize awards often lead to debates on decisions taken by the Nobel committee, even years after the fact. In this article a case study is presented on the award of the Medicine prize in 1929 to Christiaan Eijkman for his discovery that a lack of a hitherto unknown "accessory food factor" (now called vitamin B1) is the cause of beriberi. I will show, however, that it was Eijkman's successor Gerrit Grijns who was the first ever scientist to point to the "partial hunger" resulting from vitamin deficiency, an interpretation which Eijkman fiercely contested for almost thirty years. Unfortunately, Grijns' discovery was largely ignored, both by the Nobel committee and by Eijkman in his Nobel prize speech.

**Keywords:** Vitamins, beriberi, Eijkman, Grijns

## Introduction

The story is too good not to tell, and has become part of the myth of science. When working in the Dutch East Indies in the 1890s (now Indonesia), the Dutch medical officer Christiaan Eijkman (1858-1930) discovered that chicken fed with boiled, white rice developed a neurodegenerative disease that was similar to beriberi, a tropical disease that caused numerous victims at that time, both among the indigenous population and people from western colonizing countries. When fed with whole-grain rice, however, the birds remained healthy or recovered. This accidental observation would ultimately lead Eijkman to a Nobel Prize in Physiology or Medicine in 1929 "for his discovery of the anti-neuritic vitamin" or vitamin B1 as it is nowadays known. As Louis Pasteur (1822-1895) said: "Fortune favours the prepared mind." But things are never as clear cut as they may seem at first sight, and since the middle of the last century many questions have been raised regarding the background of the discovery. Initially, only Dutch scientists and science historians were able to shed some light on this matter as Eijkman and others published their findings almost exclusively in Dutch science journals, but since translations of the most important articles have become available, many others were able to "to summarize what was probably the most important work for the future of nutritional science to have been conducted anywhere in the world in the 1890s."[1]

It was Eijkman's biographer who in 1959 showed how Eijkman strongly contested the correct interpretation of his own observations and called him "the reluctant father of the vitamin theory."[2] Later Johannes Reith, professor of nutrition science and toxicology at the University of Utrecht, concluded that the contributions of Eijkman's former assistant Gerrit Grijns (1865-1944) had been overlooked completely. Moreover, according to Reith, Eijkman had given a distorted picture of the main events in the discovery process. It was Grijns who did the experiments that showed that beriberi is a deficiency disease and Grijns who coined the brilliant term "partial hunger." Eijkman fought this explanation tooth and nail for thirty years and only became convinced after the isolation of the pure vitamin B1 by Barend Coenraad Petrus Jansen (1884-1962) and Willem Frederik Donath (1889-1957) in 1926.[3] Even so, he continued to systematically downplay Grijns' role in his writings and even in his Nobel acceptance speech.[4] Whereas over the years many pieces of this vitamin B1 puzzle have come to light, there is to my knowledge no single overview that puts all available evidence into perspective, and discusses the historic route towards the discovery of vitamin B1 in its entirety. Moreover, some important documents like lecture notes from Eijkman and documents from the Nobel archives, like the considerations of the Nobel committee and the text of the nomination letters for Eijkman and others active in vitamin research, have so far not been taken into consideration. Below I will try to sketch the main developments starting in 1887, when the Dutch government sent a scientific committee to the Dutch East-Indies to try to determine the cause of beriberi, and ending in 1935, when a large group of Dutch scientists honoured Grijns for "laying the foundation of our modern science of vitamins" by publishing his groundbreaking article of 1901, in which he came to the conclusion that beriberi was due to a deficiency and coined the term "partial hunger."[5]

## A tropical disease

The earliest reference to beriberi, "a plague that makes people impotent in their hands and legs," is found in a letter from the first Dutch governor of the Dutch East-Indies, Pieter de Both (1568-1615), in 1611.[6] Later a more extensive description was given by Jacob de Bondt (1592-1631), a Dutch doctor who worked for several years on the Indonesian island of Java. He also mentions that the name of the disease derives from the Sinhalese word *beri* meaning weakness, the word being doubled meaning "a great weakness." Patients suffered from loss of feeling in hands and feet, and experienced difficulties walking and talking. Ultimately the disease led to heart failure and death.

At the end of the nineteenth century several authors noticed a link between the occurrence of the disease and the patients' diet. Notably Frederik Johannes van Leent (1830-1895), chief health officer of the Dutch Navy, concluded that beriberi was a nutrition-related disease, resulting from a deficiency of proteins and fat. He observed a clear difference in the incidence of beriberi on Dutch navy ships between native and European crew members, who were on the same ships and subject to the same hygienic conditions, the only difference between them being the rations they received. Once the rations of the native improved, the number of beriberi cases among them fell from 60 to 7%.[7] Also in Japan a young surgeon recruited by the Navy, Kanehiro Takaki (1849-1920), noticed a link between nutrition and the incidence of *kakke* as it was called. He improved the sailors' diets by including more protein and the number of cases dropped dramatically.[8]

An alarming increase under Dutch sailors and troops that had been sent to subdue an uprising in the Atjeh in Northern Sumatra, led the Dutch government in 1886 to send a committee to its colony to investigate and provide recommendations to control the disease. The committee was led by the physiologist Cornelis Pekelharing (1848-1922) and the young neurologist Cornelis Winkler (1855-1941). Before travelling to the East, they spent some time in the Berlin laboratory of Robert Koch (1843-1910). Together with Louis Pasteur, Koch had firmly founded modern bacteriology by discovering the causative agents of deadly infectious diseases like tuberculosis, cholera, and anthrax. The work of Pasteur and Koch made such a great impression that for a while it was believed that the cause of all diseases could be ascribed to microbes alone.

Thus, bacteriology became the centre and goal of medical investigations, also for Pekelharing and Winkler.[9] In Berlin they met a young doctor, Christiaan Eijkman, who was also working in Koch's laboratory. He had already spent time in the East on an earlier research mission into malaria and was therefore assigned to the committee. In this way Eijkman got to know the disease well, first in Batavia, then in Sumatra, where the Dutch troops were stationed.[10] Upon their return to the Netherlands, Pekelharing and Winkler wrote a report in which they concluded that beriberi was an infectious disease caused by a bacterium. They even tentatively identified the microorganism, even though in most cases they failed to find bacteria in the patients' blood. Therefore, more research was required, and they recommended the government set-up a local Laboratory for Bacteriology and Pathology as part of the military hospital, with Eijkman as its director.[11]

It is here that Eijkman first started working with chickens. In 1889 he noticed that after these had been on the laboratory grounds for three to four weeks, they started to suffer from a condition similar to beriberi in humans: first they were unable to stand and had trouble walking, then their breathing slowed, and they died. The affected birds all showed signs of nerve degeneration. Eijkman was not able to isolate any disease-causing bacteria from their blood and the birds became ill virtually irrespective of whether they had been injected with cultures prepared from the blood of their diseased colleagues. It looked to Eijkman like the grounds of the laboratory had been infected.[12] At that point he made his ground-breaking observation, "something that had escaped us so far." In the months that Eijkman had done his experiments, the chickens had been fed with cooked rice, leftovers from the military kitchen. A new cook had broken with this practice – as Eijkman was to say later "he refused to give military rice to civilian chickens"[13] – which had led to a disappearance of the disease from the laboratory grounds. Eijkman then started experiments feeding chickens with cooked and uncooked rice and discovered that only those that received the former got sick, while some of them recovered after a change in diet. At this point he concluded that the nerve degeneration was caused by an unknown factor present in the cooked rice. But he did not want to give up on a bacterial explanation: "Feeding with cooked rice created favourable conditions for the development of as yet unknown micro-organisms and thereby a toxic substance which led to nerve degeneration."[14]

It took almost six years before Eijkman published the results of an additional set of systematic experiments he had done.[15] He showed that the disease was not in the laboratory grounds, was unrelated to the water the rice was boiled in, did not depend on extended storage of the cooked rice or the type of rice and finally, that the rice did not contain a toxin to start with. This led him to look into the way the rice was processed. When rice is milled, first the husks are removed, leaving the red-brownish rice bran or "silver" skin. The highest quality rice was polished rice, in which all kernels are further "scoured" and then polished making them lose their outer skin and giving them a white colour. Since 1870, the use of steam-powered rice mills had made it easier to produce this high-quality "nice-looking and smooth," polished variety. Eijkman discovered that chickens that had become sick after being fed on the polished variety, completely recovered after being fed lightly processed rice with the silver skin still on. He now saw two possibilities:

> Either the skin protects the grain from the intrusion of harmful microorganisms in the intestines and thereby prevents the excessive formation of harmful products, or [it] contains substances indispensable to life and health that are absent or occur in too low concentrations in the nucleus of the rice grain.[16]

Despite numerous other experiments, chemical analyses of the various rice constituents and testing of different feeds, like low-protein tapioca and protein-rich meat, none of Eijkman's hypotheses could be fully confirmed or rejected. However, Eijkman did conclude that the common factor in all treatments had been the presence of starch, so he concluded that fermentation of starch in the chicken's crop caused some toxic effect, that could only be prevented by an unknown substance in the rice skin.[17]

## Vorderman's prison study

At this time Eijkman's health had again deteriorated, so he was about to move back to Europe. Before his departure he spoke about the results of his chicken experiments to Alphonse Vorderman (1844-1922), Inspector of the Civil Medical Service for the islands Java and Madura. Vorderman told him that he had never found beriberi among inmates of prisons where red rice was the staple diet, in contrast to prisons where polished white rice was supplied. To avoid having to rely on his memory, he there and then decided to confirm his observations by writing to all the prison medical officers under his jurisdiction, including prisons that he had never visited himself.[18] A few days before Eijkman's ship sailed, Vorderman informed him about the results of this first informal inquest: beriberi occurred in 54% of the prisons where the inmates got white rice and only in 3.7% of the prisons where red rice was the main food. Thereupon the Dutch government ordered Vorderman to visit all prisons and obtain more detailed information on the potential relationship between rice intake and the occurrence of beriberi. In doing so, he went to great lengths to prevent potential bias in his research. For instance, he did not divulge that he wanted to sample the food, to prevent the possibility that rumours about the investigation might lead the local suppliers to change the kind of rice that they provided. He obtained the statistics of beriberi relating to the year before his visit, and had several experts analyse and classify the rice samples by origin and composition, without knowing their origin.[19] The results of this meticulous study confirmed the strong link between the type of rice and occurrence of beriberi.[20]

The publication of Vorderman's study led to a heated debate in the Netherlands, in books, journals, and even newspapers. The main controversy was whether beriberi was caused by a toxic substance in people's diet, or whether nutrition was only an accessory factor for a disease that many still believed was caused by a bacterial infection.[21] Surprisingly, also the careful research of Vorderman came under attack, such that he felt the need to defend himself.[22] Because of all the controversy, the authorities in the Dutch East Indies refused to take measures to cull the disease by changing prison diets. Eijkman also seemed to have changed his views. Upon his return to the Netherlands, he had continued his experiments on polyneuritis in the Amsterdam Zoo, with mixed results: Dutch chickens did not eat white rice and only developed the characteristic symptoms after being force-fed. It made him question the effect of nutrition on beriberi and he openly attacked the views of

Van Leent and Tanaka, who could rightly be called the pioneers in the study of the relationship between beriberi and rice diets.[23] Working from the same records as Van Leent he claimed they did not support the too simple view that a deficiency of protein and fat caused the disease. With respect to Tanaka's findings, he called it "practically unthinkable that the disappearance of kakke in the Japanese Navy was predominantly due to [a change of diet]."[24] Given all the evidence, he "saw no reason to depart from the standpoint that beriberi is most probably an infection-disease."[25]

In 1898 he was appointed professor of hygiene and forensic medicine at the University of Utrecht. But whereas in his inaugural lecture he still discussed tropical diseases,[26] he would predominantly devote his research to developing methods of measuring water purity and addressing hygienic problems in his home country. However, in Batavia, Gerrit Grijns continued Eijkman's work on chickens. As a health-officer, Grijns had been Eijkman's collaborator and when Eijkman left, Grijns remained determined to elucidate the cause of beriberi. He discovered among other things that also other foodstuffs, like a local bean, called katjang hidjoe (*Phaseolus radiatus* or mung bean), were able to cure chickens that had contracted the disease after a diet of polished white rice. In a number of carefully thought-out experiments performed over the course of the next three years Grijns determined that it was neither a protein nor a mineral nor any fatty material that gave the silver skin its protective value. He also showed that Eijkman's conclusion that rice starch was required for the development of a toxic substance was incorrect. Beriberi developed independently of the presence of carbohydrates.[27] He then drew the very important conclusion that beriberi was caused by a certain "partial hunger," a deficit of an as yet unknown substance, indispensable for the health of the nervous system: "In several natural foodstuffs substances are present that cannot be missed without damage to the peripheral nervous system. [...] The isolation of these substances is complicated by the fact they are so easily disintegrated, which [...] shows that they are very complex substances. It proved impossible to replace them by simple chemical compounds."[28]

Grijns had hit the nail on the head. This was the first clear statement of what later would be called the "vitamin" concept. Unfortunately for Grijns, he only published it in Dutch, and it would not become more widely known for another 25 years. Only in 1927, at the request of the Swiss biochemist and physiologist Emil Abderhalden (1877-1950), Grijns published his version of the development of vitamin research up to that moment, including a German translation of his 1901 paper.[29] His overview ends in 1911:

> [...] the year that Funk came up with the word 'vitamine' and in which the publications on the fat-soluble factor A from Hopkins, Osborne & Mendel and McCollum and others originate. From that moment onwards the interest in the unknown food substances would become widespread.[30]

While Grijns' contribution was nominated that year, he was not considered for a Nobel Prize.

Shortly after publication of his experiments, two independent clinical trials were held in the Dutch East Indies, which confirmed the positive effect of the addition of mung beans to the regular diet of workmen at a naval harbour and patients in a mental hospital in Batavia: there were no cases of beriberi in that group, whereas in similar groups 20-40% of the patients who had either only received extra green vegetables or who had food from the grounds that had been thoroughly disinfected did contract beriberi.[31] Unfortunately, at that point the research on the disease virtually came to a halt. Vorderman died in 1902, and between 1902 and 1904 Grijns was back in the Netherlands to convalesce after he had become sick himself. Upon his return to the laboratory, he was assigned to work on other topics. However, just before he was about to embark for the Netherlands in 1902, he had received a visitor from Norway, Axel Holst (1860-1931). The Norwegian authorities had given Holst the task to elucidate the cause of a disease that affected many Norwegian sailors, then called *ship-beriberi*. Like scurvy – to which, it would ultimately turn out, it was identical– this disease was assumed to be related to an absence of fresh food and Holst wanted to start experiments like the ones that Eijkman and Grijns had done. In his first publication on this topic, five years later, he states:

> My starting point has been the excellent research studies of the Dutch authors Eijkman and Grijns on the so-called polyneuritis gallinarum. Doctor Grijns most obligingly showed me his experiments during my stay in Batavia in 1902.[32]

Together with Theodor Frølich (1870-1947), Holst would establish guinea pigs as an animal model which allowed them to systematically study the factors that led to the disease, as well as the preventive value of different substances. Grijns himself would only pick up his work on what nowadays are referred to as deficiency diseases after 1912, when he assumed the directorship of the laboratory in Batavia. He managed to get the funds for a laboratory expansion – the institute would later be named after Christiaan Eijkman – but his health again forced him to return to his home country in 1917. Three years later he was appointed to a chair at the Rijks Landbouw Hoogeschool, nowadays Wageningen University.[33]

## Other accessory factors or vitamins

There were also others who had picked up on the importance of these hitherto unknown substances for human health. In 1905, Pekelharing, former head of the committee to investigate the cause of beriberi in the Dutch East Indies, presented the results of experiments he had done in young mice. He discovered that even when feeding these with a sufficient amount of protein, fat, carbohydrates, salts, and water, they died within weeks. However, when water was replaced by milk, they remained perfectly healthy. Apparently, he concluded, "milk contains an unknown substance which is of prime importance in our nutrition, even in very small quantities."[34] Since his efforts to isolate this substance failed, he decided not to publish his results. If only he had done so, since an English colleague of Pekelharing, the biochemist Frederick Gowland Hopkins (1861-1947), had independently been doing very similar experiments. He was already well-known for having isolated the amino acid tryptophan and demonstrating its essential nature in a human diet. In 1906, he performed a series of feeding experiments using rats, which led him to suggest that in normal diets there are tiny quantities of unidentified substances that are essential for animal growth and survival.

> No animal can live upon a mixture of pure protein, fat, and carbohydrate, and even when the necessary inorganic material is carefully supplied the animal still cannot flourish. The animal body is adjusted to live either upon plant tissues or the tissues of other animals, and these contain countless substances other than the proteins, carbohydrates, and fats. [...] It is certain that there are many minor factors in all diets of which the body takes account.[35]

He called these hypothetical substances "accessory food factors" and suggested that a lack of such nutrients might lead to such diseases as scurvy and rickets. He continued his research on these accessory factors, but after six years, just like Pekelharing, he had to concede defeat being unable to isolate them, let alone determine their chemical nature.[36] And in the absence of this, no insight could be gained into their specific roles in metabolism.[37]

Independent of Hopkins two American chemists, Thomas Osborne (1859-1929) and Lafayette Mendel (1872-1935) had also started working with young rats. Initially, they contested the results of Hopkins, claiming that there were no special substances in food and that young rats can grow on artificial diets consisting of protein, fat, salts, and carbohydrates.[38] This immediately elicited a reaction from Hopkins.[39] In a letter to Osborne, he later apologized for the strong wording he had used and admitted that: "I have done so much work in this (not very successful) endeavour to separate the unknown substances which affect growth, that when your paper [...] came out I suffered from an attack of nerves."[40] One year later Osborne and Mendel discovered the growth-promoting property of butter and butterfat in young rats.[41] They were not the first to discover this as Elmer McCollum (1879-1967), working with his assistant Marguerite Davis (1887-1967), beat them by three weeks.[42] Moreover, in 1915 McCollum showed that rats even needed two different growth substances, one soluble in fats, the other in water or alcohol, which he arbitrarily labelled as growth factors A and B. This would form the basis of the later naming systems of vitamins.[43]

Subsequently, more and more researchers began actively to try to isolate these mysterious substances present in milk and rice bran, which had such a major influence on healthy growth and development. Already Eijkman and Grijns had tried to isolate the active fraction by dissolving rice bran in water or alcohol and testing extracts on birds with polyneuritis. Others determined that the sought-after compound was a relatively small molecule, and not a protein. Some (erroneously) believed that the activity arose from compounds containing phosphorus,[44] others managed to isolate an active fraction, which upon further analysis always turned out to consist of mixtures of various chemicals.[45] Working from London, Casimir Funk (1884-1967), a Polish biochemist, announced in 1912 that he had managed to isolate from rice bran the factor responsible for preventing and curing the neurodegenerative chickens' diseas. According to him this was an amine, and he even went so far as to claim that the mysterious nutrients playing a role in other deficiency diseases were also amines. He therefore coined the term "vitamine" – short for "vital amine."[46] And although he would turn out to be wrong on both counts, the name has stuck, albeit shortened to "vitamin."

In spite of all evidence that Grijns had collected, Eijkman by 1913 still continued to cling to his poison hypothesis. He considered it to be wrong "to dismiss the poison theory altogether and simply assume a direct action of the vitamin according to Grijns. In an area as complicated as this, the simplest explanation is not the most probable."[47] Moreover, he remained convinced that this poison only developed in starch, since chickens fed exclusively on sterilized meat remained healthy.[48] This elicited a reaction from Grijns, whose experiments had shown otherwise.[49] Without going into the arguments in any detail used by both, Grijns once more underlined his view that diseases like beriberi, scurvy and pellagra all seemed to be linked "to the absence of substances indispensable for life or health."[50] In the years that followed Eijkman proved utterly incapable of accepting this conclusion. When asked in a meeting of the *Dutch Society for Tropical Medicine* in 1914 to give his view on the statement that beriberi was caused by a continuous and exclusive diet of polished rice, he was not even prepared to fully dismiss a bacterial infection.[51] And six years later he stated that whereas unknown substance or substances, "which have not been obtained in a pure state," play an important role in plants and animals, "undoubtedly also other factors contribute to the disease."[52] So where Grijns consistently stood behind his original conclusion of 1901, Eijkman had been rather erratic if not confusing in the interpretation of his observations regarding polyneuritis and rice diet. Occam's razor apparently did not apply here.

Whereas Eijkman was openly unwilling to embrace the vitamin concept, Hopkins had demonstrated himself in 1906 and 1912 that certain substances he called accessory food factors were required for a healthy growth process. Since then, however, he had published very little original scientific work on vitamins except in defence of his original findings.[53] The failure to isolate any such substance was a source of frustration for him personally. That did not stop him, however, from making sure his legacy was well-protected, as Weatherall and Kamminga have shown.[54] In 1918, Hopkins had become chairman of the *Accessory Food Factors Committee*, appointed by the Medical Research Committee in the United Kingdom. Just after World War I this committee published a report "on the present state of knowledge concerning accessory food factors (vitamines)."[55] It would become a bestseller: within two years of its publication, it sold over 3,500 copies to government departments and the public. The historical introduction – written by Hopkins – was "a rhetorical masterpiece,"[56] and to the dismay of many, he all but painted himself as the discoverer of vitamins.[57] Referring to his own "classical experiments," Hopkins especially drew attention to two graphs from his 1912 paper showing "the extraordinary effect upon the growth of the experimental rats" of an "apparently insignificant addition to the diet."[58] It is the second graph which most clearly and convincingly shows the effect of small additions of milk to an artificial diet. These growth curves would later become "familiar to anyone who has ever opened a textbook of physiology."[59]

## A Nobel Prize for vitamin research

So with one self-proclaimed and one extremely reluctant discoverer of vitamins, let's see how this resulted in each person's nominations for a Nobel Prize. Together with Funk and Suzuki, Eijkman was brought forward for the first time in 1914 as the "founder of the work on vitamins,"[60] while in 1917, 1919 and 1920 he was again nominated for his work on beriberi and chicken polyneuritis.[61] In these years the examiners on behalf of the Nobel committee proved hesitant. In 1917 it was noted that "If E[ijkman] had deepened his studies in other areas and thus shown, that he realized the scope of his discovery, and if his works had been more recent, it seems to me that E[ijkman's] candidacy for the prize would be indisputable. Now I am, on the other hand, doubtful about it and thus do not propose him [for this year]." Also in 1919 the examiner remarked that Eijkman had not fully realized the meaning of his discovery "and therefore had only made a minor contribution to the completion of research in that area."[62] Even a concerted effort on behalf of his Dutch colleagues at the universities of Amsterdam and of Utrecht, with no fewer than fifteen and five nominations in 1926 and 1927, respectively, could not persuade the Nobel Committee. In these two years also Grijns was nominated, for his crucial observation that beriberi is a deficiency disease and his introduction of the term "partial hunger." Unfortunately, these would turn out to be the only nominations for a Nobel Prize he ever received.

Nominations for Hopkins only appeared from 1923 onwards, and he was consistently nominated – thirteen times in total – up to 1929 when he did receive the Prize. Here it is interesting to note that whereas nominators often referred to his work on accessory food factors, also his successful isolations of glutathione and the amino acid tryptophan were mentioned, as well as his work on lactic acid production in muscle. In 1926 and 1927 Hopkins was only nominated for this research, in 1927 even for the Chemistry Nobel, while his vitamin discovery was not mentioned at all. That year the Nobel Committee of Chemistry declined to evaluate his research, judging that its significance "pertains to the science of physiology."[63] Interestingly, upon being congratulated on his Nobel Prize, Hopkins is supposed to have muttered that he had received the Nobel Prize for the wrong reasons.[64] One can only wonder what would have happened if Hopkins had been nominated in 1926 and 1927 for his accessory food factors as in these years both Eijkman and Grijns had been nominated several times.

After 1926, the situation had taken on a totally different outlook with the breakthrough of the isolation and testing of the "anti-beriberi factor" by Jansen and Donath.[65] As we saw above, since the first attempts of Funk, scientists all over the world, including Eijkman and Grijns, had tried in vain to accomplish this. By 1916 Robert Williams, based in the Philippines had even made more than forty attempts to obtain pure vitamin from large quantities of rice polishings, but none was successful.[66] No wonder since we know now there is only about a teaspoonful of vitamin in a ton of rice bran.[67] Jansen and Donath, working from the upgraded laboratory in Batavia where also Eijkman and Grijns had done their research, succeeded in obtaining a small quantity of pure crystalline material. With tiny amounts of this – doses as small as a few micrograms – they succeeded in keeping small local birds and pigeons being fed a polished rice diet free from polyneuritis.[68] They also sent a sample to Eijkman for testing, and he confirmed their results within a few months.[69] Whether or not this may have swayed the Nobel committee and its examiners, it is a fact that Hopkins was not nominated in 1927, and Eijkman was overlooked in 1928, so it was not until 1929 that the two were jointly considered again.

That year George Barger (1878-1939), professor of medicinal chemistry at the University of Edinburgh had apparently set his sights on a Nobel Prize for vitamin research, since he nominated Hopkins for his discovery "that certain accessory factors of the diet are necessary for growth [which] laid the groundwork for the later discovery of vitamins," a nomination which "seemed to me beyond doubt." Should the committee "not be convinced of Hopkins' predominance," he suggested to either jointly award Hopkins and Eijkman, or Hopkins, Osborne and Mendel. These four names were "the only really important ones in the great field of vitamin research, and of the four that of Hopkins seems to me to be the greatest."[70] Interesting to note in this respect is Barger's reasoning for nominating Eijkman, who "interpreted the discovery of the dietetic origin of beriberi as due to a toxic substance."[71] As seen above, this was indeed Eijkman's interpretation, but it had in the meantime been shown to be fully incorrect! So if anyone truly could have stated that he received the Nobel Prize for the wrong reasons, it was Eijkman.

The 1929 report was written by the Swedish pharmacologist and secretary of the Nobel Committee, Göran Liljestrand (1886-1968). Regarding Eijkman he starts by repeating the objections of his predecessors in 1917 and 1919, and then sets out to briefly summarize Eijkman's contributions. Two things are noteworthy: Liljestrand follows Eijkman in claiming that the Vorderman study was initiated by himself, and he specifically mentions Grijns as "the first (in 1901) to pronounce the view that food together with protein, fat, carbohydrates, water and salts must also contain certain other substances to be complete," adding that "Eijkman later joined this interpretation."[72] Liljestrand then also notes that Eijkman's work had been of fundamental importance for the entire vitamin theory and that he had succeeded in introducing an effective therapy against beriberi. Whereas Eijkman's research had been done long before, according to the Nobel statutes "older works could also be considered for an award in case their importance had not been demonstrated until recently."[73] Therefore, Liljestrand recommended awarding one half of the 1929 prize to Eijkman "for his discovery of the antineuritic vitamin" and the other half to Hopkins "for his discovery of the growth stimulating vitamins." This recommendation was unanimously accepted by the Nobel Committee.

It should be noted here that Liljestrand also considered the contributions of Mendel. As described above, he had observed, in collaboration with Osborne, who had died in January that year and no longer came into consideration, that artificial diets were sufficient to promote growth in young rats. Later they had to admit that they had not worked with pure samples and had not extended their experiments over sufficiently long periods. They were able to correct these omissions, but their publication on the importance of butter fat for a healthy growth was preceded by just three weeks by McCollum and Davis. Each group went on to show convincingly the existence of a substance necessary for growth (later called vitamin A), while McCollum together with Davis also discovered another (water-soluble) substance, identical to the substance active against beriberi, later called vitamin B. Both American groups struggled to get the same clear-cut results as Hopkins had, a mystery that was never solved. Hopkins by contrast was also the only one who from the very start proclaimed his beliefs in the reality of accessory food factors in print and persisted in a search for these elusive substances. Liljestrand apparently felt safe to conclude that Mendel and Osborne's work was inferior to that of Hopkins, and the contributions of McCollum and Davis did not warrant further attention because they were not nominated that year. Therefore, they may have been the "highly qualified losers" in this Nobel vitamin tale.[74]

Because of bad health, Eijkman would never make it to Stockholm. He died a year after the ceremonies were held. In their Nobel speeches both laureates sketched their view of the development of their ideas.[75] Only one year earlier Eijkman had (finally) acknowledged the crucial role of Grijns as the one "who, well before Funk, clearly defined the idea of a deficiency disease both with respect to polyneuritis and to beriberi,"[76] but he did not mention this in his Nobel speech, in contrast to Hopkins, who did give credit where credit was due: "the view that the cortical substance in rice supplied a need rather than neutralized a poison was soon after put forward by Grijns and *ultimately* accepted by Professor Eijkman himself."[77] Hopkins also addressed the objections raised by Funk in his *Science* article of 1926.[78] While he grants that he never claimed to be the "discoverer of vitamins," he denies Funk's claim that Hopkins' paper of 1912 appeared too late to exert any influence on the development of the subject and accuses Funk of disingenuously claiming "that the substance facilitating growth found in milk is similar to, if not identical with, the vitamins described by me [Funk]." Mixing praise with critical statements, Hopkins then continues: "I would like to make clear my belief that he [Funk] has not received too much, but too little, credit for his vitamin work as a whole. I venture to think, however, that he was in no sense my predecessor in the physiological field."

## Reflection

So who really discovered vitamins? And who did or did not receive credit for this? Citing the discovery of oxygen and that of the planet Neptune in the eighteenth century, Thomas Kuhn already showed that "any attempt to date a discovery or to attribute it to an individual must inevitably be arbitrary, [...] because discovering a new sort of phenomenon is necessarily a complex process which involves recognizing both that something is and what it is."[79] That said it is true that the first one to draw the conclusion that beriberi and chicken polyneuritis were caused by a deficiency of a yet unknown compound was Gerrit Grijns, *well before Hopkins ever considered any accessory food factors.* It is also a fact that this was acknowledged by Liljestrand both in his original Nobel considerations of 1929 and later in an overview of all Nobel Medicine Prizes.[80] Whereas Grijns' article was published in a Dutch journal, and therefore did not become known to the international research community until 1935 when a translation was published, Hopkins masterfully succeeded in bringing to everyone's attention the outcome of his experiments with baby rats as due to the absence of an unknown accessory food factor.[81] Also Eijkman personally made sure that his discovery of the role of the mysterious substance in the rice silver skin was known to an international audience, while at the same time consistently denying that beriberi is a deficiency disease for almost thirty years.

Unfortunately for Grijns, he never came into consideration in 1929, *simply because in contrast to 1926 and 1927 he was not nominated.* The Nobel committee only relied upon the nomination put forward by George Barger. And he did not rank Grijns among "the four really important [names] in the great field of vitamin research," most probably because he was not aware of his research published in Dutch, and Grijns' Dutch colleagues had already nominated him earlier without success. This clearly illustrates how thin a line exists between failure and success, between oblivion and eternal fame. This story also shows the inherent arbitrariness of Nobel Prizes (and other prizes in science), for which scientific breakthroughs and discoveries of individuals are rewarded, whereas these often come about gradually, and moreover, in which numerous researchers are involved. Other factors like having a prominent position, a good network and a sense for (self-)promotion may then become crucial for receiving an award.

Only many years later, would the exlusion of Grijns be considered to be an oversight of the Nobel committee. Some noted that at the time already the award "was met with scepticism,"[82] or that "it was generally acknowledged that Grijns was entitled just as much as Eijkman and Hopkins to the Nobel Prize."[83] Others considered it "staggering to observe how someone [Eijkman in his Nobel lecture, RvdB] idealised his own work and trivialised the part of others (in this case that of Grijns) in it."[84] Grijns himself, however, never commented on or shared his disappointment with this turn of events. When he turned 70 in 1935, an international scientific committee decided to publish translations of his most important vitamin articles, in recognition for "[laying] the foundation of our modern science of vitamins."[85] Where his biographer called Eijkman "the reluctant father of vitamin theory," Gerrit Grijns is the only "true father."

**Acknowledgement**

Material from the Nobel Archives was kindly provided by the Nobel Committee for Physiology and Medicine. I would like to thank Ann-Mari Dumanski for her kind help in obtaining these documents.

## Primary sources

Barger, George, Nomination letter to the Nobel Committee for Physiology or Medicine, 1929, Nobel Archives, Stockholm.

Handwritten notes in preparation for his Nobel lecture, Noord-Hollands Archief (Haarlem), Archief van de Vereniging Nederlands Tijdschrift voor Geneeskunde (NTvG) te Amsterdam, (Verzamelde persoonlijke stukken, 5.9 Christiaan Eijkman, NL-HlmNHA_936_206).

Heubner, Wolfgang, Nomination for the Nobel Prize in Medicine (1914), Nobel Prize Nomination Archive. https://www.nobelprize.org/nomination/archive/

Holmgren, Israel, Nomination for the Nobel Prize in Medicine (1921), Nobel Prize Nomination Archive. https://www.nobelprize.org/nomination/archive/

Keibel, Franz, Nomination for the Nobel Prize in Medicine (1917), Nobel Prize Nomination Archive. https://www.nobelprize.org/nomination/archive/

Liebermann, L. von, Nomination for the Nobel Prize in Medicine (1919), Nobel Prize Nomination Archive. https://www.nobelprize.org/nomination/archive/

Liljestrand, G., Sekret Handling betänkande angående C. Eijkman, F.G. Hopkins och L.B. Mendel, 1929, Nobel Archives, Stockholm.

## Notes

1 Kenneth J. Carpenter and Barbara Sutherland, "Eijkman's contribution to the discovery of vitamins," *Journal of Nutrition* 125 (1995), 155–163.

2 B.C.P. Jansen, *Het levenswerk van Christiaan Eijkman (1858-1930)* (Haarlem: De erven F. Bohn, 1959), 117: "Men kan Eijkman zien als een soort onwillige vader van de vitaminenleer."

3 B.C.P. Jansen and W. F. Donath, "Over het isoleeren van het antiberi-beri vitamine," *Geneeskundig Tijdschrift voor Nederlandsch-Indië* 66 (1926), 573–574.

4 J.F. Reith, "Christiaan Eijkman en Gerrit Grijns," *Voeding* 32 (1971), 180–195.

5 *Prof. Dr. G. Grijns researches on vitamins 1900-1911 and his thesis on the physiology of the nervus opticus translated and reedited by a committee of honour on occasion of his 70th birthday* (Gorinchem: J. Noorduyn en Zoon, 1935), XVI.

6 M.A. van Andel, "Bontius tropische geneeskunde. Inleiding," in *Opuscula Selecta Neerlandicorum de Arte Medica* (Amsterdam: Amstelodami, 1931), XXX.

7 F.J. Van Leent, "Communication sur le béri-béri," *Congrès Périodique International des Sciences Médicales* 6 (1880), 170–94, on 184–185.
F.J. Van Leent, "Mededeelingen over beri-beri," *Geneeskundig Tijdschrift voor Nederlandsch Indië* 20 (1880), 272–310.

8 Yoshinori Itokawa, "Kanehiro Takaki (1849-1920). A biographical sketch," *Journal of Nutrition* 106 (1976), 582–588.

9 K. Codell Carter, "The germ theory, beri-beri, and the deficiency theory of disease," *Medical History* 21 (1977), 119–136.

10 B.C.P. Jansen, *Het levenswerk*, 11–12.

11 C.A. Pekelharing and C. Winkler, *Onderzoek naar den aard en de oorzaak der beri-beri en de middelen om die ziekte te bestrijden* (Utrecht: Kemink en Zoon: Utrecht, 1888).

12 C. Eijkman, "Polyneuritis bij hoenderen," *Geneeskundig Tijdschrift voor Nederlandsch Indië* 30 (1890), 295–334. Eijkman's reports on his findings with chickens, which were originally published in the *Geneeskundig Tijdschrift voor Nederlandsch-Indië*, 30 (1890), 295–334; 32 (1892), 353–362 and 36 (1896), 214–269, have been republished in an English translation by D.G. van der Heij in C. Eijkman, *Polyneuritis in chickens, or the origin of vitamin research*, (Basel: Hoffman-La Roche, 1990).

13 C. Eijkman, "Anti-neuritis-vitamine en beriberi. Nobelprijs-voordracht 1929," *Nederlands Tijdschrift voor Geneeskunde* 134 (1990), 1654–1657, on 1656.

14 C. Eijkman, Polyneuritis bij hoenderen, 328.

15 C. Eijkman, "Polyneuritis bij hoenderen. Nieuwe bijdrage tot de aetiologie der ziekte," *Geneeskundig Tijdschrift voor Nederlandsch-Indië* 36 (1896), 214–269.

16 Ibid., 229.

17 Ibid., 265–267.

18 A.G. Vorderman, *Onderzoek naar het verband tusschen den aard der rijstvoeding in de gevangenissen op Java en Madoera en het voorkomen van beri-beri onder de geïnterneerden*, (Batavia: Jav. Boekh. & Drukkerij, 1896), 1–2.

19 Jan Vandenbroucke, "Adolphe Vorderman's 1897 Study on Beriberi: An Example of Scrupulous Efforts to Avoid Bias," *Journal of the Royal Society of Medicine* 106 (2013), 108–111.

20 Vorderman, *Onderzoek naar het verband*, 58–59 and Bijlage 10.

21 Jaap Bos, *Evart van Dieren. Een kroniek van het falen* (Amsterdam: Amsterdam University Press, 2008), 113–121.

22 A.G. Vorderman, "Toelichting op mijn beri-beri verslag," *Geneeskundig Tijdschrift voor Nederlandsch-Indië* 38 (1898), 47–68.

23 Harmke Kamminga, "Credit and Resistance: Eijkman and the Transformation of Beri-beri into a Vitamin Deficiency Disease," in *From Physico-Theology To Technology: Essays in The Social and Cultural History of Biosciences: A Festschrift for Mikulas Teich*, eds. Kurt Bayertz and Roy Porter (Amsterdam-Atlanta: Rodopi, 1998), 232–254.

24 C. Eijkman, "Beri-beri en voeding: een kritisch historische studie," *Nederlandsch Tijdschrift voor Geneeskunde* 34 (1898), 186–209, 233–247 and 275–303, on 243.

25 Ibid., 289.

26 Christiaan Eijkman, *Over gezondheid en ziekte in heete gewesten*, (J. van Druten, Utrecht, 1898).

27 G. Grijns, "Over polyneuritis gallinarum," *Geneeskundig Tijdschrift voor Nederlandsch-Indië* 41 (1901), 3–110, on 31.

28 Ibid., 48–49.

29 G. Grijns, "Beiträge zur Geschichte der Erkennung der Beriberi als Avitaminose," *Fortschritte der Naturwissenschaften. Neue Folge* 1 (1927), 1–32.

30 Ibid., 32.

31 M.D.J. Hulshoff Pol, "Katjang idjo un nouveau medicament contre le beri-beri," *Janus* 7 (1902), 524–535 and 570–581.

32 Axel Holst, "Experimental Studies Relating to Ship Beri-Beri and Scurvy. Introduction," *Journal of Hygiene* 7 (1907), 619–633 and 623–624.

33 Marinus C. Kik, "Gerrit Grijns (1865-1944)," *Journal of Nutrition* 62 (1957), 2–12.

34 C.A. Pekelharing, "Over onze kennis van de waarde der voedingsmiddelen uit chemische fabrieken," *Nederlandsch Tijdschrift voor Geneeskunde* 41 (tweede reeks) (1905), 111–124, on 122.

35 F. Gowland Hopkins, "The Analyst and the Medical Man," *The Analyst* 31 (1906), 385–404, on 395.

36 F. Gowland Hopkins, "Feeding experiments illustrating the importance of accessory factors in normal dietaries," *Journal of Physiology* 44 (1912), 425–460.

37 Harmke Kamminga and Mark Weatherall, "The Making of a Biochemist I: Frederick Gowland Hopkins' Construction of Dynamic Biochemistry," *Medical History* 40 (1996), 269–292; Mark Weatherall and Harmke Kamminga, "The Making of a Biochemist II: The Construction of Frederick Gowland Hopkins' Reputation," *Medical History* 40 (1996), 415–436.

38 Thomas B. Osborne and Lafayette B. Mendel, "Feeding experiments with fat-free food mixtures," *Journal of Biological Chemistry* 12 (1912), 81–89; Thomas B. Osborne, Lafayette B. Mendel, and L. Ferry Edna, "Beobachtungen über Wachstum bei Fütterungsversuchen mit isolierten Nahrungssubstanzen," *Hoppe Seyler's Zeitschrift für physiologische Chemie* 80 (1912), 307–370; Thomas B. Osborne, Lafayette B. Mendel, and L. Ferry Edna, "Maintenance experiments with isolated proteins," *Journal of Biological Chemistry* 13 (1912), 233–276.

39 Frederick Gowland Hopkins and Allan Neville, "A note concerning the influence of diets upon growth," *Biochemical Journal* 7 (1913), 97–99.

40 As quoted in Stanley Becker, "Will Milk Make Them Grow? An Episode in the Discovery of the Vitamins," in *Chemistry and modern society: historical essays in honor of Aaron J. Ihde*, eds. John Parascandola and James C. Whorton (American Chemical Society Print Book, Washington DC, 1983), 61–83, on 68.

41 Thomas B. Osborne and Lafayette B. Mendel, "The relation of growth to the chemical constituents of a diet," *Journal of Biological Chemistry* 15 (1913), 311–326.

42 Elmer V. McCollum and Marguerite Davis, "The necessity of certain lipins in the diet during growth," *Journal of Biological Chemistry* 15 (1913), 167–175.

43 Elmer V. McCollum and Cornelia Kennedy, "The dietary factors operating in the production of polyneuritis," *Journal of Biological Chemistry* 24 (1916), 491–502.

44 H. Schaumann, "Further Contributions to the Etiology of Beri-beri," *Transactions of the Society of Tropical Medicine and Hygiene* 5 (1911), 59–75; H. Schaumann, "Beriberi und Nucleinphosphorsäure in der Nahrung," *Archiv für Schiffs- und Tropen-Hygiene* 12 (Beiheft 5) (1908), 37–57.

45 U. Suzuki, T. Shimamura and S. Odake, "Über Oryzanin, ein Bestandteil der Reisklei und seine physiologische Bedeutung," *Biochemische Zeitschrift* 43 (1912), 89–153; E.B. Vedder and R.R. Williams, "Concerning the Beriberi-Preventing Substances or Vitamines Contained in Rice Polishings," *Philippine Journal of Science* 8B (1913), 175–195.

46 Casimir Funk, "The Etiology of Deficiency Diseases," *The Journal of State Medicine (1912-1937)* 20, Vol. 6 (1912), 341–368, on 342.

47 C. Eijkman, "Aetiologie und Prophylaxis der Beri-Beri," *Transactions of the 17th International Congress of Medicine London 1913. Section 21: Tropical medicine and hygiene* (London: Oxford University Press, 1913), 25–40, on 33. Interestingly, in a virtually identical Dutch translation of his talk at the International Congress of Medicine, Eijkman does not mention that Vorderman started his study at Eijkman's instigation, even though this is written in his lecture notes. In these notes he also specifically states that Grijns' explanation is too simple, something which is again absent in the printed version. See: C. Eijkman, "Aetiologie en prophylaxis der beri-beri," *Nederlandsch Tijdschrift voor Geneeskunde* 57, Vol 2 (1913), 1426–1438. Eijkman's handwritten notes in preparation for this lecture can be found in the Noord-Hollands Archief (Haarlem), Archief van de Vereniging Nederlands Tijdschrift voor Geneeskunde (NTvG) te Amsterdam, (Verzamelde persoonlijke stukken, 5.9 Christiaan Eijkman, NL-HlmNHA_936_206). C. Eijkman, "Über die Natur und Wirkungsweise der gegen experimentelle Polyneuritis wirksamen Substanzen," *Archiv für Schiffs- und Tropen-Hygiene* 17 (1913), 328–335.

48 C. Eijkman, "Über Ernährungspolyneuritis," *Archiv für Hygiene* 58 (1906), 150–170.

49 G. Grijns, "Über Ernährungspolyneuritis. Abwehr gegen Prof. Dr. C. Eijkman's Kritik," *Archiv für Hygiene* 62 (1907), 128–136.

50 Ibid., 136.

51 "Vereenigingsverslagen. Nederlandsche Vereeniging voor Tropische Geneeskunde," *Nederlandsch Tijdschrift voor Geneeskunde* 58, Vol.1 (1914), 684–688, on 688.

52 B. Brouwer, "Vergadering Amsterdamsche Neurologen-Vereeniging," *Nederlandsch Tijdschrift voor Geneeskunde* 64, Vol. 2 (1920), 359–362, on 361.

53 Frederick Gowland Hopkins and Allen Neville, "A Note Concerning the Influence of Diets upon Growth," *Biochemical Journal* 7 (1913), 97–99; Frederick Gowland Hopkins, "Note on the Vitamine Content of Milk," *Biochemical Journal* 14 (1920), 721–724.

54 Weatherall and Kamminga, "The Making of a Biochemist II."

55 *A Report on the Present State of Knowledge Concerning Accessory Food Factors (Vitamines). Compiled by a Committee Appointed Jointly by the Lister Institute and the Medical Research Committee* (London: H.M. Stationery Office, 1919).

56 Weatherall and Kamminga, "The Making of a Biochemist II," 418.

57 Several years later Casimir Funk would express his indignation about this state-of-affairs in a short article in the journal *Science*, in which he claimed that Hopkins' work came years after the work of the "real" pioneers, like Eijkman, Grijns, Schaumann, Suzuki, and obviously, himself. See: Casimir Funk, "Who Discovered Vitamins," *Science* 63 (1926), 455–456.

58 *A Report on the Present State of Knowledge*, 9–10.

59 W.R. Aykroyd, *Vitamins and Other Dietary Essentials* (London: Heinemann (Medical Books) Ltd, 1933), 46.

60 Nomination by Wolfgang Heubner in 1914, Nobel Prize Nomination Archive. https://www.nobelprize.org/nomination/archive/

61 Nominations by Franz Keibel, L. von Liebermann and Israel Holmgren in 1917, 1919 and 1920, respectively, Nobel Prize Nomination Archive. https://www.nobelprize.org/nomination/archive/

62 O. Sundberg, *Yearbook 1917*, as quoted in G. Liljestrand, Sekret Handling betänkande angående C. Eijkman, F.G. Hopkins och L.B. Mendel, 1929, Nobel Archives, Stockholm.

63 As quoted in Weatherall and Kamminga, "The Making of a Biochemist II," 416.

64 Rolf Zetterström, "C. Eijkman (1858-1930) and Sir F.G. Hopkins (1861-1947): The Dawn of Vitamins and other Essential Nutritional Growth Factors," *Acta Pœdiatrica*, 95 (2006), 1331–1333, on 1333.

65 B.C.P. Jansen and W.F. Donath, "Antineuritisch vitamine," *Chemisch Weekblad* 23 (1926), 201–203. B.C.P. Jansen and W.F. Donath, "Over de isoleering van het anti-beriberi-vitamine," *Verslagen van de Koninklijke Akademie van Wetenschappen (Wis- en natuurkundige afdeling)* 35 (1926), 923–934.

66 Robert R. Williams, "The Chemistry of Vitamins," *The Philippine Journal of Science* 11 (1916), 49–57.

67 Kenneth J. Carpenter, *Beriberi, White Rice, and Vitamin B. A Disease, a Cause, and a Cure*, (Berkeley:University of California Press, 2000), 110.

68 B.C.P. Jansen and W.F. Donath, "Over de isoleering van het antiberiberi-vitamine," *Geneeskundig Tijdschrift voor Nederlandsch-Indië* 66 (1926), 810–827, on 821–822.

69 C. Eijkman, "Proeven met het antiberi-beri vitamine van Jansen en Donath," *Verslagen van de Koninklijke Akademie van Wetenschappen (Wis- en natuurkundige afdeling)* 36 (1927), 221–228; C. Eijkman, "Experiments with Jansen and Donath's Antiberiberi-Vitamin," *Proceedings Royal Academy of Sciences* 30 (1927), 376–382.

70 Nomination letter of George Barger to the Nobel Committee for Physiology or Medicine, 1929, Nobel Archives, Stockholm.

71 Ibid. Osborne received another nomination from J. Le Roy Conel "for his work on preparing chemically standardized diets and analyzing the effects of diets having certain vitamin deficiencies when fed to rats," and Hopkins was also nominated by the Edinburgh physiologist Sir Edward

Sharpey Schafer for his "Discovery of 'accessory factors' (now: vitamins), as essential nutritional components, Isolation of tryptophan and glutathione."

72 G. Liljestrand, Sekret Handling betänkande angående C. Eijkman, F.G. Hopkins och L.B. Mendel, 1929, Nobel Archives, Stockholm, 3–4.

73 Ibid., 5.

74 F. Luttenberger, "Excellence and Chance: the Nobel Prize Case of E. von Behring and É. Roux," *History and Philosophy of the Life Sciences*, 18 (1996), 225–238.

75 Due to ill health, Eijkman was not able to come to Stockholm to deliver his Nobel Lecture and sent the text for inclusion in *Les Prix Nobel en 1929* (Stockholm: Imprimerie Royale. P. A. Norstedt & Söner, 1930).

76 C. Eijkman, "Zur Vorgeschichte der Vitaminlehre," *Karlsbader ärztliche Vorträge* 8 (1928), 137–160; C. Eijkman, "Over de voorgeschiedenis der vitamineleer," *Nederlandsch Tijdschrift voor Geneeskunde* 72 (1928), 5571–5578: "Ook is het Grijns geweest, die nog vóór Funk het begrip deficiëntieziekte ten aanzien zoowel van de experimenteele polyneuritis als van de beri-beri scherp geformuleerd heeft."

77 Sir Frederick Hopkins, "The Earlier History of Vitamin Research," Nobel Lecture, December 11, 1929. In *Nobel Lectures, Physiology or Medicine 1922-1941* (Amsterdam; Elsevier Publishing Company, 1965) (italics added); Christiaan Eijkman, "Anti neuritic Vitamin and Beriberi," in *Nobel Lectures, Physiology or Medicine 1922-1941* (Amsterdam: Elsevier Publishing Company, 1965).

78 Casimir Funk, "Who discovered vitamins," *Science* 63 (1926), 455–456.

79 Thomas Kuhn, "Historical Structure of Scientific Discovery," *Science*, 136 (1962), 760–764, on 762.

80 Göran Liljestrand, "The Prizes in Physiology or Medicine," in *Nobel. The man and his Prizes*, (Amsterdam, London, New York: Elsevier Publishing Company, 1962), 131–343, on 238.

81 Cf. in a 1922 exchange of letters in *Nature*, an anonymous editor even refers to the unknown "anti-rachitic factor" (later called vitamin A) as "Hopkins' stuff." See: W.M. Bayliss, "The Cause of Rickets," *Nature* 110 (1922), 212.

82 J. Verhoef, H. Snippe and H.S.L.M. Nottet, "Christiaan Eijkman. First bacteriologist at Utrecht University, Nobel laureate for his work on vitamins," *Antonie van Leeuwenhoek* 75 (1999), 165–169.

83 S. Postmus, "Gerrit Grijns 1865-1944," *Voeding*, 16 (1955), 3–4.

84 Reith, "Christiaan Eijkman en Gerrit Grijns," 192.

85 *Prof. Dr. G. Grijns' Researches on Vitamins, 1900-1911* (Gorinchem; J. Noorduyn en Zoon, 1935), v.

## Bibliography

Andel, M.A. van. "Bontius tropische geneeskunde. Inleiding," in *Opuscula Selecta Neerlandicorum de Arte Medica.* Amsterdam: Amstelodami 1931): xxx.

Anonymous. "Vereenigingsverslagen. Nederlandsche Vereeniging voor Tropische Geneeskunde," *Nederlandsch Tijdschrift voor Geneeskunde* 58 (1914): 684–688.

Anonymous. *A Report on the Present State of Knowledge Concerning Accessory Food Factors (Vitamines). Compiled by a Committee Appointed Jointly by the Lister Institute and the Medical Research Committee.* London: H.M. Stationery Office, 1919.

Aykroyd, W.R. *Vitamins and Other Dietary Essentials.* London: Heinemann (Medical Books) Ltd, 1933.

Bayliss, W.M. "The Cause of Rickets," *Nature* 110 (1922): 212.

Becker, Stanley. "Will Milk Make Them Grow? An Episode in the Discovery of the Vitamins." In *Chemistry and Modern Society: Historical Essays in Honor of Aaron J. Ihde*, edited by John Parascandola and James C. Whorton, 81–83. Washington DC: American Chemical Society Print Book, 1983.

Bos, Jaap and Evart van Dieren. *Een kroniek van het falen*. Amsterdam: Amsterdam University Press, 2008.

Brouwer, B. "Vergadering Amsterdamsche Neurologen-Vereeniging," *Nederlandsch Tijdschrift voor Geneeskunde* 64 (1920): 359–362.

Carpenter, Kenneth J. and Barbara Sutherland. "Eijkman's Contribution to the Discovery of Vitamins," *Journal of Nutrition* 125 (1995): 155–163.

Carpenter, Kenneth J. *Beriberi, White Rice, and Vitamin B. A Disease, a Cause, and a Cure*. Berkeley: University of California Press, 2000.

Carter, K. Codell. "The Germ Theory, Beri-Beri, and the Deficiency Theory of Disease," *Medical History* 21 (1977): 119–136.

Eijkman, C. "Polyneuritis bij hoenderen," *Geneeskundig Tijdschrift voor Nederlandsch-Indië* 30 (1890): 295–334.

Eijkman, C. "Polyneuritis bij hoenderen. Nieuwe bijdrage tot de aetiologie der ziekte," *Geneeskundig Tijdschrift voor Nederlandsch-Indië* 36 (1896): 214–269.

Eijkman, C. "Beri-beri en voeding: een kritisch historische studie," *Nederlandsch Tijdschrift voor Geneeskunde* 34 (1898): 186–209, 233–247 and 275–303.

Eijkman, Christiaan. *Over gezondheid en ziekte in heete gewesten*. Utrecht: J. van Druten, 1898.

Eijkman, C. "Über Ernährungspolyneuritis," *Archiv für Hygiene* 58 (1906): 150–170.

Eijkman, C. "Aetiologie en prophylaxis der beri-beri," *Nederlandsch Tijdschrift voor Geneeskunde* 57 (1913): 1426–1438.

Eijkman, C. "Aetiologie und Prophylaxis der Beri-Beri," in *Transactions of the 17th International Congress of Medicine London 1913*. Section 21: Tropical medicine and hygiene, 25–40. London: Oxford University Press, 1913.

Eijkman, C. "Über die Natur und Wirkungsweise der gegen experimentelle Polyneuritis wirksamen Substanzen," *Archiv für Schiffs- und Tropen-Hygiene* 17 (1913): 328–335.

Eijkman, C. "Experiments with Jansen and Donath's Antiberiberi-Vitamin," *Proceedings Royal Academy of Sciences* 30 (1927): 376–382.

Eijkman, C. "Proeven met het antiberi-beri vitamine van Jansen en Donath," *Verslagen van de Koninklijke Akademie van Wetenschappen* (Wis- en natuurkundige afdeling) 36 (1927): 221–228.

Eijkman, C. "Over de voorgeschiedenis der vitamineleer," *Nederlandsch Tijdschrift voor Geneeskunde* 72 (1928): 5571–5578.

Eijkman, C. "Zur Vorgeschichte der Vitaminlehre," *Karlsbader ärztliche Vorträge* 8 (1928): 137–160.

Eijkman, C. "Anti-neuritis-vitamine en beriberi. Nobelprijs-voordracht 1929," *Nederlands Tijdschrift voor Geneeskunde 134* (1990): 1654–1657 and 1656.

Eijkman, C. *Polyneuritis in Chickens, or the Origin of Vitamin Research*. Basel: Hoffman-La Roche, 1990.

Eijkman, Christiaan. "Anti neuritic Vitamin and Beriberi," in *Nobel Lectures, Physiology or Medicine 1922-1941*. Amsterdam: Elsevier Publishing Company, 1965.

Funk, Casimir. "The Etiology of Deficiency Diseases," *The Journal of State Medicine* 20 (1912): 341–368.

Funk, Casimir. "Who Discovered Vitamins," *Science* 63 (1926): 455–456.

Grijns, G."Over polyneuritis gallinarum," *Geneeskundig Tijdschrift voor Nederlandsch-Indië* 41 (1901): 3–110.

Grijns, G. "Über Ernährungspolyneuritis. Abwehr gegen Prof. Dr. C. Eijkman's Kritik," *Archiv für Hygiene* 62 (1907): 128–136.

Grijns, G. "Beiträge zur Geschichte der Erkennung der Beriberi als Avitaminose," *Fortschritte der Naturwissenschaften*. Neue Folge 1 (1927): 1–32.

Grijns, G. *Prof. Dr. G. Grijns' Researches on Vitamins 1900-1911 and his Thesis on the Physiology of the Nervus Opticus*, translated and reedited by a committee of honour on occasion of his 70th birthday. Gorinchem: J. Noorduyn en Zoon, 1935.

Holst, Axel. "Experimental Studies Relating to Ship Beri-Beri and Scurvy. Introduction," *Journal of Hygiene* 7 (1907): 619–633 and 623–624.

Hopkins, F. Gowland. "The Analyst and the Medical Man," *The Analyst* 31 (1906): 385–404.

Hopkins, F. Gowland. "Feeding Experiments Illustrating the Importance of Accessory Factors in Normal Dietaries," *Journal of Physiology* 44 (1912): 425–460.

Hopkins, Frederick Gowland, and Allen Neville. "A Note Concerning the Influence of Diets upon Growth," *Biochemical Journal* 7 (1913): 97–99.

Hopkins, Frederick Gowland. "Note on the Vitamine Content of Milk," *Biochemical Journal* 14 (1920): 721–724.

Hopkins, Sir Frederick. "The Earlier History of Vitamin Research." Nobel Lecture, December 11, 1929. In *Nobel Lectures, Physiology or Medicine 1922-1941*. Amsterdam: Elsevier Publishing Company, 1965.

Hulshoff Pol, M.D.J. "Katjang idjo un nouveau medicament contre le beri-beri," *Janus* 7 (1902): 524–535 and 570–581.

Itokawa, Yoshinori. "Kanehiro Takaki (1849-1920). A Biographical Sketch," *Journal of Nutrition* 106 (1976): 582–588.

Jansen, B.C.P., and W.F. Donath. "Antineuritisch vitamine," *Chemisch Weekblad* 23 (1926): 201–203.

Jansen B.C.P. and W. F. Donath. "Over het isoleeren van het antiberi-beri vitamine," *Geneeskundig Tijdschrift voor Nederlandsch-Indië* 66 (1926): 573–574.

Jansen, B.C.P. and W.F. Donath. "Over de isoleering van het antiberiberi-vitamine," *Geneeskundig Tijdschrift voor Nederlandsch-Indië* 66 (1926): 810–827.

Jansen, B.C.P. and W.F. Donath. "Over de isoleering van het anti-beriberi-vitamine," *Verslagen van de Koninklijke Akademie van Wetenschappen* (Wis- en natuurkundige afdeling) 35 (1926): 923–934.

Jansen, B.C.P. *Het levenswerk van Christiaan Eijkman (1858-1930)*. Haarlem: De erven F. Bohn, 1959.

Kamminga, Harmke and Mark Weatherall. "The Making of a Biochemist I: Frederick Gowland Hopkins' Construction of Dynamic Biochemistry," *Medical History* 40 (1996): 269–292.

Kamminga, Harmke. "Credit and Resistance: Eijkman and the Transformation of Beri-Beri into a Vitamin Deficiency Disease," in *From Physico-Theology To Technology: Essays in The Social and Cultural History of Biosciences: A Festschrift for Mikulas Teich*, edited by Kurt Bayertz and Roy Porter, 232–254. Amsterdam, Atlanta: Rodopi, 1998.

Kik, Marinus C. "Gerrit Grijns (1865-1944)," *Journal of Nutrition* 62 (1957): 2–12.

Kuhn, Thomas. "Historical Structure of Scientific Discovery," *Science* 136 (1962): 760–764.

Leent, F.J. van. "Communication sur le béri-béri," *Congrès Périodique International des Sciences Médicales* 6 (1880): 170–94.

Leent, F.J. van. "Mededeelingen over beri-beri," *Geneeskundig Tijdschrift voor Nederlandsch-Indië* 20 (1880): 272–310.

Liljestrand, Göran. "The Prizes in Physiology or Medicine," in *Nobel: The Man and his Prizes*, edited by Henrik Schück, 131–343. Amsterdam, London, New York: Elsevier Publishing Company, 1962.

Luttenberger, Franz. "Excellence and Chance: the Nobel Prize Case of E. von Behring and É. Roux," *History and Philosophy of the Life Sciences*, 18 (1996): 225–238.

McCollum, Elmer V. and Marguerite Davis. "The Necessity of Certain Lipins in the Diet during Growth," *Journal of Biological Chemistry* 15 (1913): 167–175.

McCollum, Elmer V. and Cornelia Kennedy. "The Dietary Factors Operating in the Production of Polyneuritis," *Journal of Biological Chemistry* 24 (1916): 491–502.

Osborne, Thomas B. and Lafayette B. Mendel. "Feeding Experiments with Fat-Free Food Mixtures," *Journal of Biological Chemistry* 12 (1912): 81–89.

Osborne, Thomas B., Lafayette B. Mendel and L. Ferry Edna. "Maintenance Experiments with Isolated Proteins," *Journal of Biological Chemistry* 13 (1912): 233–276.

Osborne, Thomas B., Lafayette B. Mendel and L. Ferry Edna. "Beobachtungen über Wachstum bei Fütterungsversuchen mit isolierten Nahrungssubstanzen,“ *Hoppe Seyler's Zeitschrift für physiologische Chemie* 80 (1912): 307–370.

Osborne, Thomas B. and Lafayette B. Mendel. "The Relation of Growth to the Chemical Constituents of a Diet," *Journal of Biological Chemistry* 15 (1913): 311–326.

Pekelharing, C.A. and C. Winkler. *Onderzoek naar den aard en de oorzaak der beri-beri en de middelen om die ziekte te bestrijden.* Utrecht: Kemink en Zoon: Utrecht, 1888.

Pekelharing, C.A. "Over onze kennis van de waarde der voedingsmiddelen uit chemische fabrieken," *Nederlandsch Tijdschrift voor Geneeskunde* 41 (series 2) (1905): 111–124.

Postmus, S. "Gerrit Grijns 1865-1944," *Voeding* 16 (1955): 3–4.

Reith, J.F. "Christiaan Eijkman en Gerrit Grijns," *Voeding* 32 (1971): 180–195.

Schaumann, H. "Beriberi und Nucleinphosphorsäure in der Nahrung,“ *Archiv für Schiffs- und Tropen-Hygiene* 12 (Beiheft 5) (1908): 37–57.

Schaumann, H. "Further Contributions to the Etiology of Beri-Beri," *Transactions of the Society of Tropical Medicine and Hygiene* 5 (1911): 59–75.

Suzuki, U., T. Shimamura and S. Odake. "Über Oryzanin, ein Bestandteil der Reisklei und seine physiologische Bedeutung,“ *Biochemische Zeitschrift* 43 (1912): 89–153.

Vandenbroucke, Jan. "Adolphe Vorderman's 1897 Study on Beriberi: An Example of Scrupulous Efforts to Avoid Bias," *Journal of the Royal Society of Medicine* 106 (2013): 108–111.

Vedder, E.B. and R.R. Williams. "Concerning the Beriberi-Preventing Substances or Vitamines Contained in Rice Polishings," *Philippine Journal of Science* 8B (1913): 175–195.

Verhoef, J., H. Snippe and H.S.L.M. Nottet. "Christiaan Eijkman. First bacteriologist at Utrecht University, Nobel laureate for his work on vitamins," *Antonie van Leeuwenhoek* 75 (1999): 165–169.

Vorderman, A.G. *Onderzoek naar het verband tusschen den aard der rijstvoeding in de gevangenissen op Java en Madoera en het voorkomen van beri-beri onder de geïnterneerden.* Batavia: Jav. Boekh. & Drukkerij, 1896.

Vorderman, A.G."Toelichting op mijn beri-beri verslag," *Geneeskundig Tijdschrift voor Nederlandsch-Indië* 38 (1898): 47–68.

Weatherall, Mark. and Harmke Kamminga, "The Making of a Biochemist II: The Construction of Frederick Gowland Hopkins' Reputation," *Medical History* 40 (1996): 415–436.

Williams, Robert R. "The Chemistry of Vitamins," *The Philippine Journal of Science* 11 (1916): 49–57.

Zetterström, Rolf. "C. Eijkman (1858-1930) and Sir F.G. Hopkins (1861-1947): The Dawn of Vitamins and other Essential Nutritional Growth Factors," *Acta Pædiatrica* 95 (2006): 1331–1333.

CHAPTER NINE

# Konrad Lorenz, Nicolaas Tinbergen, and Karl von Frisch: the Scientific Network and the Controversy over the Nobel Prize in Physiology or Medicine in 1973

*Daniela Angetter-Pfeiffer*

**Abstract**

The chapter deals with the knowledge transfer between Nobel laureates Konrad Lorenz and Nikolaas Tinbergen. Since then, both have had a significant impact on studies of instinctive behavior, indeed in very different ways. However, their different ways of working could be perfectly combined. The article also sheds new light on the general differences and similarities between the researchers, especially regarding their social, scientific, and ethnic backgrounds, especially in relation to the Second World War and the Nazi regime.

**Keywords:** Konrad Lorenz, Nikolaas Tinbergen, Karl von Frisch, National Socialism, Careerist and exhorter

In 1973, the Austrians Konrad Lorenz (1903–1989)[1] and Karl von Frisch[2] (until 1918 Karl Ritter von Frisch, 1886–1982[3]), together with the Dutch-born Nikolaas Tinbergen (1907–1988)[4], were awarded the Nobel Prize in Physiology or Medicine for work on the organization and triggering of individual and social behaviour patterns. All three made seminal contributions to ethology, now called animal behaviour, and for a long time they were the only Nobel laureates in the field.[5] It would take nearly 50 years before another Nobel Prize was awarded for research on evolutionary history, Svante Pääbo's prize for "decoding the Neanderthal genome" in 2022.[6]

In his speech, Professor Börje Cronholm (1913–1983) of the Karolinska Institute emphasized that the Nobel Prize laureates of 1973 had collected a great deal of data on animal behaviour in natural environments but also in experimental situations and studied the functions of behavioural patterns, the role in the individual struggle for survival and for the continued existence of the species. Thus, behavioural patterns crystallized as results of natural selection as well as morphological traits and physiological functions. Of fundamental insight is that some behavioural patterns are obviously genetically programmed, require no prior experience, and are automatically triggered by certain key stimuli. They function mechanically, practically robotically, and are no longer influenced by external circumstances after the triggering stimulus. Concerning insects, fish, and birds, for example, courtship, nesting and brood care are among these fixed patterns of action. Because the discoveries were based on studies of insects, fish, and birds, they initially seemed to be of only minor importance for human physiology or medicine.[7]

Konrad Lorenz had first observed and studied his fundamental research on imprinting on jackdaws, primarily on his own-hand-raised jackdaw "Tschok", which Lorenz interpreted as its "mother" and which, even as an adult, refused to break away from him and live together with its conspecifics. But other jackdaws that Lorenz raised on his own also refused to leave him later.[8] Lorenz was not the first to study the phenomenon of imprinting, but he was the first to intuitively grasp connections, to recognize the importance of this behavior, and to know how to combine theory and practice. In addition, he recast the aggression drive and coined such terms as infantile schema and key stimulus.[9]

Karl von Frisch's achievements in the fields of comparative physiology and behavioural research included work on the sensory physiology of honeybees, like the perception that when bees find nectar in a flower, they fly in a special pattern and perform a kind of "dance" that shows other bees where to find the nectar.[10]

Nikolaas Tinbergen studied, among other research questions, the function of key stimuli and innate trigger mechanisms on insects, in the brood-care behaviour of herring gulls, and in the courtship chain of sticklebacks.[11] With his book "The study of instinct," published in 1951, he wrote the first textbook of comparative behavioural research. He considered it a particular success of his work that he had succeeded in "spreading the ethological principles in the English-speaking world."[12,13]

In any case, the Karolinska Institute in Stockholm justified the award of the Nobel Prize in Physiology or Medicine by stating that the discoveries of the three scientists had stimulated extensive research in the field of mammals, for example on monkeys, and were subsequently also of great importance for psychiatry and research into psychosomatic diseases.[14] Thus, studies are devoted, first, to the existence of genetically programmed behavioural patterns, their organization, maturation, and their triggering by key stimuli, and, second, to the significance of specific experiences in critical phases for the normal development of an individual. Psychosocial disorders can lead not only to behavioural disorders but also to such severe somatic diseases as arterial hypertension and myocardial infarction.[15] The Nobel Prize to Lorenz, Frisch, and Tinbergen thus elevated the study of animal behaviour to the status of serious biological research.[16] And this experimental elucidation of the neural basis of biological cognition systems belonged to the main interest of a branch of science that began to develop especially since the last decade, that of neuro-ethology.[17]

None of the three scientists would have thought that the award of the Nobel Prize in Physiology or Medicine would be overshadowed by Konrad Lorenz' National Socialist past, which was to come to the attention of general public in 1973, not only in Austria but also internationally.[18] Lorenz was regarded as an avowed National Socialist, while Tinbergen and Frisch were opponents of the Nazi regime. In recent biographical research on Frisch, however, it is critically noted that, in the summer of 1944, he advocated eugenics in his publications.[19]

In addition to the scientific collaboration between the three researchers, this article therefore focuses on the debate surrounding Lorenz' Nazi past and the award of the Nobel Prize, especially since Konrad Lorenz is today considered one of the most internationally renowned Austrians. He was awarded no fewer than ten honorary doctorates for his achievements in behavioural research, and was admitted to prestigious scientific societies after World War II, including the Austrian Academy of Sciences in 1951, the German Academy of Natural Scientists Leopoldina in 1957, the Royal Society as a foreign member in 1964, and the National Academy of Sciences of the USA in 1966. Likewise, he received countless awards and quite a few institutes and schools in Austria and Germany are named after him. In Austria, but also internationally, Lorenz' name is also inseparably linked with the Konrad Lorenz Institute of the Austrian Academy of Sciences (today the Konrad Lorenz Institute for Comparative Behavioral Research of the University of Veterinary Medicine) as well as with the anti-nuclear movement concerning the establishment of a national park instead of a power plant in Hainburg, Lower Austria.[20] According to the Austrian sociologist and science journalist Klaus Taschwer, there is "hardly any other Austrian scientist after 1945 who would have received an ever-increasing number of honours during his lifetime and who would have been as well-known internationally."[21]

## Konrad Lorenz – a Nobel hero in the shadow of the Nazi past

When the phone call came from Stockholm on October 12, 1973, to tell Konrad Lorenz that he had just been awarded the Nobel Prize in Physiology or Medicine, his wife Margarete (1854–1986) took the call. A short time later, confirmation came from a neighbour who had heard the announcement on the radio,[22] and then also by telegram, including a very unusual typo: the Altenburg postmistress had turned the Nobel Prize into a "Nopelpreis."[23] In a first reaction Lorenz said: "Of course I am enormously pleased about this prize, but not so much because it brings me personal recognition. It is more important that obviously also in the general public the knowledge about behavioral research is becoming more and more widespread, from that research which I have always carried out on animals and which is of the greatest importance for humans as such. From all the findings, inferences for humanitas are obvious."[24]

The call from Stockholm was a surprise to Karl von Frisch, who was 87 years old at that time. In a first reaction, Frisch said "I am delighted about the award, also for my homeland Austria!"[25] Later he added that it filled him with pride to be honoured together with Lorenz and Tinbergen.[26]

Austria rejoiced, because for the first time since the end of World War II, this highest scientific award in the world went to the Republic again.[27]

At almost the same time in Oxford, Nikolaas Tinbergen received the message from his secretary. She had taken the phone call from Stockholm and knew immediately what was going on.[28,29]

For Tinbergen and Frisch, the award must have come as a genuine surprise, as Tinbergen's student and later biographer Hans Kruuk described in his biography about Tinbergen in 2003: They had not expected ethology to become so established in science.[30] Lorenz had expected it sooner. He had already been considered as a candidate previously.[31,32] In any case, the three quickly became "Heroes of the Year" according to Kruuk.

In a letter to the zoologist Otto Koehler (1889–1974), who had congratulated him, Konrad Lorenz summed up precisely the tripartite nature of the prize: "I alone would probably not have been worthy of it, and Niko alone perhaps not either, but together we are, or rather it is the result of what we have done together. Karl von Frisch, of course, deserves the Nobel Prize long ago, and it can only deeply honour Niko and me to be held in the same esteem to this truly great man."[33]

Hardly anyone suspected at the time that Lorenz would so quickly be caught in the crossfire of public and media criticism in view of his Nazi past, such that he would even be forced to renounce acceptance of the Nobel Prize.

To understand the discussion whether Lorenz was a "worthy" Nobel laureate, some milestones of the three behavioural scientists are outlined here very briefly.

The fact that Konrad Lorenz and Niko Tinbergen were in the foreground as representatives of behavioural research was summarized by Kruuk in 2003 as follows: "It is difficult to realize how far we have travelled from the early days of studying animal behaviour. Before Niko arrived on the scene, behaviour science was focussed largely on white rats and pigeons behind bars. Things that happened out in the wild were rarely respectable subjects for scientific enquiry. Now we see this fabulous richness of displays, gestures, attacks, and courtship in all creatures around us, and we simply cannot imagine not asking questions about that. Much of this change is due to 'ethology', the discipline of Konrad Lorenz and Niko Tinbergen."[34] Karl von Frisch found no mention in this quote. In any case, Tinbergen, who also spoke excellent German, influenced mainly the Anglo-American world, while Lorenz became the leading figure of behavioural research in the German-speaking world.[35]

Tinbergen's contact with Lorenz took place in the 1930s. When Tinbergen, after his return from Greenland in 1933, was commissioned to design a new practical course concerning experiments with the behaviour of selected animals and was instructed to give accompanying lectures, he became aware of Lorenz' publications on jackdaws, especially the article published in the "Journal für Ornithologie" in 1935, "Der Kumpan in der Umwelt des Vogels. Der Artgenosse als auslösendes Moment sozialer Verhaltungsweisen." The publication, which had abruptly made Lorenz internationally famous, dealt with the fact that species-specific behavioural patterns can serve as taxonomic characteristics just as much as organs, and had already become the basis for the interpretation of observable behaviour in Leiden immediately after its publication. Lorenz' article was in any case the reason for Tinbergen to start a correspondence with the Austrian private scholar. When Lorenz wrote Tinbergen about a planned trip to Belgium, Tinbergen persuaded his institute director, Cornelis Jakon van der Klaauw (1893–1972), to invite Lorenz to a symposium in Leiden in November 1936 on the topic of instinct. Tinbergen tailored this event precisely to Lorenz' scientific focus.[36] Lorenz accepted the invitation, and Tinbergen now got to know the Viennese behavioural scientist personally. The two understood right away and recognized a strong common research interest. In 1985, Tinbergen reflected on Lorenz's visit to Leiden at that time: "The theoretically strong but experimentally weak guest was enthusiastic about the experiments with sticklebacks running in the host's laboratory: 'This is exactly what I need.'"[37] Lorenz commented in 1937 in a letter to the German zoologist Oskar Heinroth (1871–1945) that Tinbergen was a terribly nice and quite excellently clever fellow.[38] In any case, this personal meeting was the beginning of a fruitful scientific collaboration and lifelong friendship, which, however, was to be repeatedly put to the test due to the Third Reich and Lorenz's advocacy of racial ideological and eugenic measures. First, however, the meeting in Leiden induced Lorenz to invite Tinbergen to Altenberg with his wife and their son. On Lorenz's family estate, Tinbergen had had the opportunity to raise and observe birds.[39] Tinbergen, together with his family and a basket full of teal, pintail, and gadwall ducks, came to stay with Lorenz for three months in March 1937. Planned for this stay in Lower Austria were: Exchange experiences, collect ideas and discuss theories, conduct observations and experiments in Lorenz's private zoo. The main focus was on a core problem of the young subject of ethology: what is innate and what is learned?[40]

Lorenz tended to link his observations to basic evolutionary and epistemological concepts. Tinbergen, by contrast, linked his observations to experiments that demonstrated methodological precision and analytical clarity. Whereas Lorenz was often enticed by vision and enthusiasm, Tinbergen demanded that every idea had to be rigorously tested by experiment. His experimental design was always simple and clearly structured, and Tinbergen, as experimenter, succeeded in confirming

Lorenz's excellently formulated explanatory models concerning animal behavior. These different approaches proved to be a perfect combination during the common research.[41]

Together, Lorenz and Tinbergen analyzed the phenomenon of imprinting in goslings and published the study "Taxis und Instinkthandlung in der Eirollbewegung der Graugans" which was still current decades later and served as material for school lessons. It was published in 1938 in the second issue of the "Zeitschrift für Tierpsychologie"[42] (Journal of Animal Psychology, from 1986 "Ethology"), which had been founded a year earlier and was the world's first professional journal in this field.[43] Tinbergen transferred the instinct terminology developed therein into a hierarchical instinct model in 1951. But this instinct theory of Lorenz and Tinbergen was subjected to a comprehensive description, analysis, and harsh criticism from the 1980s by Hanna-Maria Zippelius, who worked as a scientific assistant for Karl von Frisch from 1943 to 1945 on the control of bee epidemics. Based on experiments she doubted that Tinbergen had correctly interpreted the mechanisms of egg recognition in herring gulls,[44] and also refuted that herring gulls have no innate knowledge of their parents.[45] Many of his findings were later thought to be wrong. Even his psychohydraulic model that instinct actions are triggered by key stimuli had become outdated. Zippelius was not the only one who criticized Lorenz and Tinbergen. Positively Zippelius did amit, however, that without Lorenz and Tinbergen there would be no behavioral research.[46]

In addition, investigations into reactions to overflying dummy birds of prey or geese were carried out in Altenberg[47] and Tinbergen also completed his experiments on the locking (begging) of young blackbirds and song thrushes, which he had started in Leiden.[48] Konrad Lorenz later described the collaboration as follows: "This summer with Niko Tinbergen was the very best of my life"[49] and Tinbergen characterized the visit to Altenberg as the "fairy tale" of his life.[50] After Tinbergen had finished his research stay with Konrad Lorenz in Altenberg, he personally contacted Karl von Frisch and accepted his invitation to his Munich institute. Tinbergen had been influenced by Frisch even earlier than by Konrad Lorenz, since Tinbergen had worked with Frisch's methods for his research on the beewolf project, from which his dissertation resulted.[51]

## The break caused by World War II

After 1937, Lorenz and Tinbergen parted ways, but there was still a continuous written correspondence and a constant scientific exchange. However, this communication ended abruptly with the outbreak of World War II. Tinbergen was interned for almost two years from October 1942 in the NS camp Beekvliet in St. Michielsgestel because he had demonstrated together with members of the teaching staff against the dismissal of Jewish colleagues at Leiden University and was also active as an underground resistance fighter.[52] In a letter, Tinbergen later described the situation in the following terms: "Our university was, by accident, the first group of Dutchmen to be tackled by the Germans as a group, and the first to refuse to surrender. The Germans wanted to 'cleanse' our corps of Jews and anti-Nazis and proceeded to fire one professor, then another, step by step, on wholly irrelevant grounds. Soon we saw no other way than to resist by refusing to stay in the service of the German-controlled government, and soon after the University was closed by the Germans because of anti-German 'irregularities' sixty of our professors including myself laid down their functions. This was at the same time our protest and our means to prevent the Germans to nazify the University."[53] Although the German zoologist and behavioural researcher Otto Koehler as well as Konrad Lorenz immediately offered their help and wanted to try to get Tinbergen released from the camp with their political contacts. However, Nikolaas and his wife Lies Tinbergen, firmly refused this assistance. In May 1946, Tinbergen justified this decision in a letter to Koehler: Out of solidarity with his fellow prisoners, such special treatment would have been out of question. Konrad Lorenz was taken as prisoner of war by the Russians in 1944. The relationship between the two scientists was severely clouded due to their different attitudes towards the Nazi regime at the time. This was certainly also due to the fact that Tinbergen had not only experienced camp imprisonment, but also witnessed deportations of his Jewish neighbors, but also recognized the danger his family was exposed to because they hid Jews.[54] His war trauma that developed from all these experiences led to depression and prevented him from accepting a position as department director at the Max Planck Institute in 1959. Neither he nor his wife would feel completely at home in Germany, despite many good friends there. But he certainly would not want to live with some Germans. And he further clarified this decision: "Completely irrational, a stupid trauma – but it doesn't work."[55]

In 1948, after Lorenz had returned to Austria from captivity, he contacted Niko Tinbergen by letter. A year later, the two met at an international meeting of ethologists in Cambridge.[56] There was a lengthy discussion about her wartime past. Afterwards, Lorenz then summed up: "Although he [Tinbergen] had spent a few years in a German concentration camp[57] and I had spent a long time in Soviet prisoner-of-war camps, we found that this had not changed anything in any way." After the conference, Lorenz spent another ten days in Leiden. Further meetings followed, for example, in Wilhelmshaven in 1950, in Buldern in 1952, organized by Lorenz and Tinbergen, and in Starnberg in 1961.[58]

## Discussion around Konrad Lorenz's Nobel Prize

Shortly after the announcement of the laureates, on October 22, 1973, media in Germany and Austria quoted passages from Lorenz's publication "Durch Domestikation verursachte Störungen arteigenen Verhaltens" which was published in 1940.[59] In this article, as well as in "Die angeborenen Formen möglicher Erfahrungen" published in the *Zeitschrift für Tierpsychologie* in 1943, some statements corresponded entirely to the programme and rhetoric of National Socialism. The first-mentioned article contained, among other messages, the following statement:

> Das immer von neuem mögliche Auftreten von Menschen mit Ausfällen im arteigenen sozialen Verhalten bildet eine Schädigung für Volk und Rasse, die schwerer ist als die einer Durchmischung mit fremdrassigen, denn diese ist wenigstens als solche erkennbar und nach einmaliger züchterischer Ausschaltung nicht weiter zu fürchten. Sollte sich [...] herausstellen, daß unter den Bedingungen der Domestikation [...] der Wegfall der natürlichen Auslese [...] die Unausgeglichenheit der Stämme verschuldet, so müßte die Rassenpflege dennoch auf schärfere Ausmerzung ethisch Minderwertiger bedacht sein.[60]

> The ever new possible occurrence of humans with deficiencies in the social behavior of the species constitutes a damage to the people and the race, which is more severe than that of an intermixture with foreign breeds, because this is at least recognizable as such and after a one-time elimination by breeding is no longer to be feared. If it should turn out [...] that under the conditions of domestication [...] the omission of natural selection [...] is to blame for the imbalance of the tribes, then the care of the races would nevertheless have to be intent on a stricter elimination of ethically inferior ones [Translation by Angetter].

Such statements had already been criticized in 1955 by the geneticist John Burdon Sanderson Haldane (1892–1964). However, it did not have a lasting impact.[61] And for the Nobel Committee Lorenz's formulations apparently were deemed irrelevant. One has to assume that people in Stockholm were familiar with Lorenz's publications, especially since his works were frequently cited in the international scientific community.[62] It was not until the 1970s that science historians began to research Lorenz's Nazi past more intensively, documented his party membership and analyzed his publications with regard to Nazi terminology.[63] Lorenz's National Socialist convictions were already clearly evident in 1937 and he received funding and support from the Notgemeinschaft der Deutschen Wissenschaft, a forerunner of today's Deutsche Forschungsgemeinschaft. It is true that his first application in 1937 was rejected because his parentage and political views were questioned. Only the second application, supported by letters of recommendation attesting to his Aryan ancestry and confirming that he was politically impeccable in every respect and that his biological research would be of beneficial to the German Reich, was successful.[64]

Lorenz was enthusiastic about progress and, like many biologists of the time, sympathized with the idea of eugenics. He also hoped – but without success – that political opportunism would help him to establish his own research institute.[65] In his application for admission to the NSDAP of June 28, 1938, he wrote:

> Ich war als Deutschdenkender und Naturwissenschaftler selbstverständlich immer Nationalsozialist und aus weltanschaulichen Gründen erbitterter Feind des schwarzen Regimes (nie gespendet oder geflaggt) und hatte wegen dieser auch aus meinen Arbeiten hervorgehenden Einstellung Schwierigkeiten mit der Erlangung der Dozentur. Ich habe unter Wissenschaftlern und vor allem Studenten eine wirklich erfolgreiche Werbetätigkeit entfaltet, schon lange vor dem Umbruch war es mir gelungen, sozialistischen Studenten die biologische Unmöglichkeit des Marxismus zu beweisen und sie zum Nationalsozialismus zu bekehren. Schließlich darf ich wohl sagen, daß meine ganze wissenschaftliche Lebensarbeit, in der stammesgeschichtliche, rassenkundliche und sozialpsychologische Fragen im Vordergrund stehen, im Dienste nationalsozialistischen Denkens steht.[66]

> As a German thinker and natural scientist, I was, of course, always a National Socialist and, for ideological reasons, a bitter enemy of the black regime (never donated or flagged) and because of this attitude, which is also evident from my work, I had difficulties in obtaining a lectureship. I have had, among scientists and especially students, a very successful campaign unfold, long before the upheaval I had succeeded in showing socialist students the biological impossibility of Marxism and in converting them to National Socialism. Finally, I may say that my entire life's scientific work, in which questions of phylogenetic, racial history and social-psychological questions are in the foreground, is in the service of National Socialism thinking [Translation by Angetter].

His involvement in the Office of Racial Policy [Rassenpolitisches Amt] is also undisputed.[67]

But even after 1945, Lorenz used a language that borrowed heavily on Nazi ideology. In his 1973 publication "Acht Todsünden der zivilisierten Menschheit", for example, he referred to people who, in his opinion, had not matured "norms of social behaviour". He described them as "parasites of society" and compared them to "malignant tumours". For him, the increasing juvenile delinquency at that time was a sign of "genetic deterioration" and in 1988 he still had "a certain sympathy for HIV" as a "means" against the worldwide "overpopulation".[68] In 1973 he formulated the following statement:

> Das verderbliche Wachstum bösartiger Tumoren beruht, wie schon angedeutet, darauf, daß gewisse Abwehrmaßnahmen versagen oder von den Tumorzellen unwirksam gemacht werden, mittels deren der Körper sich sonst gegen das Auftreten ‚asozialer' Zellen schützt. Nur wenn diese vom umgebenden Gewebe als seinesgleichen behandelt

> und ernährt werden, kann es zu dem tödlichen infiltrativen Wachstum der Geschwulst kommen. Die schon besprochene Analogie lässt sich hier weiterführen. Ein Mensch, der durch das Ausbleiben der Reifung sozialer Verhaltensnormen in einem infantilen Zustand verbleibt, wird notwendigerweise zum Parasiten der Gesellschaft. Er erwartet als selbstverständlich die Fürsorge der Erwachsenen weiter zu genießen, die nur dem Kinde zusteht. […] Wenn die fortschreitende Infantilisierung und wachsende Jugend-Kriminalität des Zivilisationsmenschen tatsächlich, wie ich befürchte, auf genetischen Verfallserscheinungen beruht, so sind wir in schwerster Gefahr.[69]
>
> The fatal growth of malignant tumours is based, as already indicated, on the fact that certain defence measures fail or are made ineffective by the tumour cells, by means of which the body otherwise protects itself against the appearance of "asocial" cells. Only when these are treated and nourished by the surrounding tissue as equals can the lethal infiltrative growth of the tumour occur. The analogy already discussed can be continued here. A person who remains in an infantile state due to the failure of social norms of behaviour to mature necessarily becomes a parasite of society. He expects, as a matter of course, to continue to enjoy the care of adults, which is due only to the child. […] If the progressive infantilization and growing juvenile delinquency of civilized man is indeed, as I fear, due to genetic decay, we are in great danger [Translated Angetter].

And only a few pages earlier in this publication he clarified his views with the words:

> Am entgegengesetzten, den das Pendel vor nicht allzu langer Zeit durchlaufen hat, stehen Eichmann und Auschwitz, stehen Euthanasie, Rassenhaß, Völkermord und Lynchjustiz. Wir müssen uns klar darüber werden, daß zu beiden Seiten des Punktes, auf den das Pendel wiese, wenn es je zur Ruhe käme, echte Werte stehen: auf der »linken« der Wert der freien individuellen Entfaltung, auf der »rechten« Seite der Wert der sozialen und kulturellen Gesundheit.[70]
>
> At the other end, which the pendulum passed not so long ago, stand Eichmann and Auschwitz, stand euthanasia, racial hatred, genocide, and lynch laws. We must realize that on either side of the point where the pendulum would refer, if it ever came to rest, real values appear: on the "left" the value of free individual development, on the "right" side the value of social and cultural health [Translated Angetter].

On the occasion of the awarding the Nobel Prize, it was therefore not surprising that heated discussions as to whether Lorenz was even worthy of receiving the award occurred. Lorenz was explicitly asked from various sides to distance himself from the writings he had published not only during the Nazi era but also later. In a first reaction to the accusations in the media, Lorenz was intransigent. He emphasized to the Austrian daily newspaper "Kurier" that he was embarrassed about the matter, but he never saw himself in a political role and condemned his critics with harsh words: "Jeder, der mich in die Nähe der Nazi stellen will, ist eine Dreckschleuder."[71] [Anyone who wants to put me in the vicinity to the Nazis is a mud slinger, translated Angetter]. In his interview he denied the fact that he had possessed the party book of the NSDAP.

Nikolaas Tinbergen also wrote to Lorenz forcing him to distance himself from his statements made during the National Socialist regime.[72] Lorenz did not even even think it worth the trouble to reply to this letter. On November 8, 1973, Lorenz gave an interview to a Dutch television crew in his Austrian hometown Altenburg, in which he denied having been a member of the Nazi Party but was firm in saying that there was inferior human material for him. To the journalist Vic Cox Lorenz expressed his opinion in the following way:

> Ich glaubte natürlich, daß von den Nazis vielleicht etwas Gutes kommen könnte. Viel bessere, intelligentere Menschen als ich, darunter mein Vater, glaubten das. Daß die Nazis Mord meinten, als sie „Auswahl" sagten, kam keinem in den Sinn. Ich glaubte niemals an die Nazi-Ideologie, dachte aber wie ein Narr, ich könnte diese Menschen bessern, sie zu etwas Besserem führen. Es war ein naiver Irrtum.[73]

> Of course, I believed that something good might come from the Nazis. Far better, more intelligent people than me, including my father, believed that. It never occurred to anyone that the Nazis meant murder when they said "selection." I never believed in Nazi ideology but thought like a fool that I could improve these people, lead them to something better. It was a naive mistake [Translated Angetter].

The interview was broadcast on November 26, 1973, by the Nederlands Christelijke Radio Vereniging in the series "Hier en nu". Statements by Lorenz were compared with statements by Tinbergen, who was also interviewed for this programmme.[74] Possibly this interview was one reason why the Austrian-Jewish architect and writer Simon Wiesenthal, who had founded the Jewish Historical Documentation Centre in Linz in 1947 and later the Documentation Centre of the Association of Jewish Victims of the Nazi Regime in Vienna, appealed to Lorenz in an open letter on November 22, 1973, not to accept the Nobel Prize and formulated:

> Der Nobelpreis darf durch die Verleihung an Sie, der einst zu den Thesen einer unbarmherzigen Diktatur stand, nicht entwertet werden.[75]

> The Nobel Prize must not be devalued by awarding it to you who once accepted the theses of a ruthless dictatorship [Translated Angetter].

Again, there was no response from Lorenz. On December 7, shortly before he left for Stockholm for the award ceremony, Lorenz issued a statement to the Austrian press agency APA, in which he expressed some regret, but again focused on the fact that he had been misunderstood and that some of his remarks had been badly distorted. In Stockholm, he and his family had received a very warm welcome, although a large number of journalists asked questions about his Nazi past, in particular about his NSDAP membership and about his racial psychological research commissioned by the National Socialists. At the Nobel Prize ceremony in Stockholm, Lorenz had personal protection, and at the press conference together with Nikolaas Tinbergen he officially regretted and that he was sorry to have seen positive aspects of National Socialism; he now sees many things differently. In this situation, he even received support from Tinbergen, who particularly emphasized in this press conference that the friendship and the fruitful cooperation had already begun before the Nazi regime. Discussions about Lorenz's Nazi past did not go any deeper in Stockholm and even the radical leftist newspaper reporters gradually stopped their hostility and started to behave more calmly. In addition, Lorenz was advised, as he told, "from the highest authority" not to make any more statements about his past, because "every explanation, if someone wants to be malicious, is twisted into its opposite and has only negative effects because it is interpreted as an excuse."[76]

In an interview for Konrad Lorenz's 85th birthday with "Der Spiegel," the subject of his Nazi was raised again. Lorenz answered the journalist's question "As the son of a wealthy physician, you were born with a golden spoon in your mouth. Didn't that shape you in your social attitude, for example, deprive you of the opportunity to imagine how the very poor are doing?" with the following statement:

> Sicher. Mit den ganz Armen habe ich wenig Mitleid gehabt. Und ich habe mich ja auch vor aller Politik gedrückt, weil ich mit meinen Problemen beschäftigt war. Auch vor einer Auseinandersetzung mit den Nazis habe ich mich in sehr verächtlicher Weise gedrückt, ich hatte einfach keine Zeit dazu." Auf die Reaktion des Journalisten, dass ihm deswegen später Vorwürfe gemacht worden sind, meinte Lorenz: „Ich mache mir selber Vorwürfe. Andererseits: Wenn ich mich frühzeitig meiner politischen Pflichten erinnert hätte, hätte ich viele Dinge, für die ich den Nobelpreis bekommen habe, nie geschaffen.[77]

> Sure. I had little sympathy for the very poor. And I also avoided all politics because I was occupied with my own problems. I also shied away from a confrontation with the Nazis very contemptuously, I simply didn't have time for it." In response to the journalist's reaction that he was later accused of this, Lorenz said: "I reproach myself. Nevertheless, if I had remembered my political duties early on, I would never have achieved many of the things for which I received the Nobel Prize [Translation by Angetter].

– A statement that official Austria had advocated years earlier.

Karl von Frisch did not attend the Nobel Prize ceremony, but he was represented by his son, the zoologist Otto von Frisch (1929–2008).[78]

If Frisch is seen by the U.S. historian of science Tania Munz as "staunchly apolitical"[79], Klaus Taschwer believed he was a supporter of Nazi eugenics. Frisch's book "Du und das Leben," published in 1936, initially also advocated the "Gesetz zur Verhütung erbkranken Nachwuchses und für das Rassenpolitische Amt"[80] (Law for the Prevention of Hereditarily Disabled Offspring), which came into force in 1934.[81] In this bestseller, Frisch formulated resolutely that if man tries to carry out an elimination of the unfit, this will be done by human methods. Sterilization is a safe means of preventing the reproduction of a hereditary disease, because marriage counseling and marriage bans are not sufficient as prevention, since love does not like to be dictated to. This passage was deleted in later editions.[82]

## Austria's dealing with the Nazi past

It would be a long time before the Republic of Austria acknowledged the deeds committed during the Third Reich. Although denazification measures were taken after the end of World War II, there was a long phase of the reintegration of National Socialists and a dominance of the war generation began as early as 1946. Until the 1970s, discussing Austria's complicity in World War II was largely taboo.[83]

It was not until the mid-1980s, not least due to international pressure and under the aspect of the Waldheim affair, an internationally conducted debate about the suspected involvement of the candidate for the office of Federal President Kurt Waldheim (1918–2007) in war crimes during the National Socialist era, that Austria began to increasingly critically look at its own history, in both the political and academic spheres. On the one hand, new restitution laws and reparation payments were passed and all victims of National Socialism were recognized; on the other hand, the medical faculty of the University of Vienna (now Medical University of Vienna) also increasingly began to confront itself with its darkest point of history starting in the 1990s. But despite all this research and its consequences like renaming of streets, withdrawal of awards and so forth, many questions are still open and still represent important research desiderata today.[84]

In any case, Konrad Lorenz's attitude toward the Third Reich has had repercussions up to the recent present. In 2015, the University of Salzburg revoked his honorary doctorate.[85] This revocation aroused great media interest in Austria and even led to a parliamentary question in January 2016. It said: The accusation that Konrad Lorenz approved of National Socialist ideas weighs heavily. Especially since it can be assumed that the awarding of his honorary doctorate, and even more so the Nobel Prize, was preceded by a thorough examination of the candidate, which not only included his scientific achievements, but also attached importance to an impeccable curriculum vitae. The withdrawal of the honorary doctorate, seen in this light, is tantamount to the accusation that generations of honourable reviewers, first and foremost the Nobel Prize Committee, had made a careless decision and that the countless honorary doctorates that Konrad Lorenz was awarded, including from renowned universities such as Basel, Yale, and Oxford, were due to a single error.[86] In view of the discussion surrounding the award of the Nobel Prize, the jury had to be aware of Lorenz's Nazi past. The University of Salzburg justified the withdrawal by saying that Lorenz had concealed his Nazi past in the honorary procedure and was therefore not worthy of the award.[87]

The assumption is therefore obvious that in the context of honours, awards and also the Nobel Prize, Konrad Lorenz's personality was never judged in its entirety, but by his scientific work. This weighs particularly heavily, since Lorenz did not distance himself from his racial ideological and eugenic views even after 1945, but continued to use this term until the 1970s, even though his Nazi past had no explicit impact on his career after 1945. Konrad Lorenz is one of those Austrians who continued to be active in the scientific community after 1945 and which means that Austria's process of coming to terms with the past has not yet been completed to this day.

## Notes

1 Konrad Zacharias Lorenz, born in Vienna in 1903 as the son of the well-known orthopaedist Adolf Lorenz (1854–1946), grew up in Altenberg in Lower Austria, studied medicine in New York from 1922, and at the University of Vienna from 1923. Earned a Doctor of medicine in 1928. Initially an assistant at the II. Anatomical Institute of the University of Vienna, he also studied zoology from 1928. In 1933 he earned a PhD, 1937 habilitation for zoology with special regard to comparative anatomy and animal psychology, 1940 becamefull professor and head of the Institute for Comparative Psychology at the University of Königsberg in East Prussia. During World War II, Lorenz was assigned first as a motorcyclist and then as an army psychiatrist and neurologist in a military hospital in Posen, where he participated in racial "studies" of "German-Polish hybrids" in 1942. From 1944–1948 he was held in Russian captivity. In 1950 Lorenz followed a call to Bristol, in 1951 he became director of the Research Center for Behavioral Physiology in Schloss Buldern in Münsterland, in 1953 he was an honorary professor at the University of Münster. In 1954, he was given a department of the Max Planck Institute for Behavioral Psychology, which was in Seewiesen in Upper Bavaria from 1956. Became honorary professor at the University of Munich in 1957, the official inauguration of the Institute of Behavioural Physiology in Seewiesen took place in 1958, and Lorenz served as its director from 1961 until his retirement in 1973. In 1974, he became an honorary professor of comparative behaviour at the University of Vienna – See Daniela Angetter, *Die österreichischen Medizinnobelpreisträger* (Österreichisches Biographisches Lexikon) Schriftenreihe 8 (Wien: Verlag der österreichischen Akademie der Wissenschaften, 2003) 58–70; Giorgio Celli, *Konrad Lorenz. Begründer der Ethologie* (Heidelberg: Spektrum der Wissenschaft, 2001).

2 Karl Ritter von Frisch was born in Vienna in 1886, as son of Anton Ritter von Frisch (1859–1917), a university professor of surgery and urology. From 1905, he studied medicine at the University of Vienna, but soon switched to zoology and received his doctorate in Vienna in 1910. In the same year he got an assistant position in Munich. In 1912, private lecturer in zoology and comparative anatomy at the University of Munich, in 1921, he became full professor of zoology and director of the Zoological Institute in Rostock, he came to Breslau in 1923 and returned to Munich in 1925. Forced to retire during the Nazi era as a "Mischling Zweiten Grades," he was recalled to the Zoological Institute because of a bee epidemic. Due to a bomb attack on the institute, Frisch moved his place of work to Brunnwinkl in the Salzkammergut region. Even after World War II, he initially remained in Brunnwinkl. In 1946, he received the chair of zoology at the University of Graz. In 1950, he returned to the reestablished Munich institute. Emeritus in 1958, he remained honorary professor at the University of Graz until 1960. Angetter, *Die österreichischen Medizinnobelpreisträger*, 71–88. When Frisch left Graz in 1950, he tried to make Lorenz his successor. Despite Lorenz' being ranked first several times for the succession, he failed due to the then Minister of Education Felix Hurdes. Although from today's perspective one might suspect Lorenz' National Socialist past as the reason for this, it was proclaimed at the time that Lorenz' penchant for Darwinism was the reason for the non-appointment, an argument that corresponded to the way formerly incriminated National Socialists were dealt with in Austria at the time. – See Benedikt Föger and Klaus Taschwer, *Die andere Seites des Spiegels. Konrad Lorenz und der Nationalsozialismus* (Wien: Czernin Verlag, 2001), 191–192.

3 The "Adelsaufhebungsgesetz" regulates the abolition of the nobility after the Habsburg monarchy during the transition to a republican form of government. It was passed on April 3, 1919, by the parliament of the newly created state Deutsch-Österreich, the Constituent National Assembly, and came into force on April 10, 1919.

4 Nikolaas (Niko) Tinbergen was born in The Hague in 1907, as son of a teacher and author. From 1925, he studied biology at Leiden University, where the zoologist Jan Verwey (1899–1981) encouraged Tinbergen's interest in ornithology. In 1930, Tinbergen took an assistant position there at the Zoological Institute; he earned a Dr. phil. in 1932. After a scientific expedition to Greenland in 1932–1933, he was again an assistant in Leiden before becoming professor of experimental zoology at the university there in 1940. In 1938, he spent several months in the USA. In 1947, he was appointed professor of experimental zoology at Leiden University. In 1949, he moved to Oxford, where he established the Research Department of Animal Behaviour, which very quickly became a hub for international behavioural biologists alongside Lorenz' Research Institute in Seewiesen. Appointed full professor of animal behaviour in the Zoological Institute at Oxford University in 1966, Tinbergen saw himself as a potential exporter of Lorenzian – essentially Austrian, Dutch, and Swiss – ideas to the English-speaking world. Emeritus in 1974. – See Hans Kruuk, *Niko's Nature. The Life of Niko Tinbergen and his Science of Animal Behaviour* (Oxford: Oxford University Press, 2003); Nikolaas Tinbergen, "Watching and wondering," in *Studying animal behavior. Autobiographies of the Founders*, edited by Donald A. Dewsbury (Chicago/London: Chicago University Press, 1985), 449.

5 Karl Schulze-Hagen, "Tinbergen und deutsche Ornithologen: Eine wechselseitige Inspiration," *Vogelwarte* 59 (2001), 7.

6 Max-Planck-Gesellschaft, " Nobelpreis 2022 für Svante Pääbo," https://www.mpg.de/19316167/nobelpreis-fuer-svante-paeaebo (accessed March 31, 2023).

7 Börje Cronholm, "Award ceremony speech," https://www.nobelprize.org/prizes/medicine/1973/ceremony-speech/ (accessed March 31, 2023).

8 Celli, *Konrad Lorenz*, 48–49.

9 Celli, *Konrad Lorenz*, 48–49.

10 Nomination Archive, "Karl von Frisch Facts," https://www.nobelprize.org/prizes/medicine/1973/frisch/facts/ (accessed February 14, 2023).

11 Schulze-Hagen, "Tinbergen," 7.

12 Tinbergen, "Watching and wondering," 453.

13 These differed significantly from the methodology of the American school of behaviourists, which had previously been the sole dominant school, which primarily conducted laboratory studies on a few selected animal species, mostly rats and pigeons, under strictly standardized conditions and nevertheless derived generally valid theories of behaviour from them. In this way, they differed fundamentally from the European school, which was designed to observe animals in the wild.

14 Nomination Archive, "Karl von Frisch Facts," https://www.nobelprize.org/prizes/medicine/1973/frisch/facts/ (accessed February 14, 2023); Nomination Archive, "Konrad Lorenz Facts," https://www.nobelprize.org/prizes/medicine/1973/lorenz/facts/ (accessed February 14, 2023); Nomination Archive, "Nikolaas Tinbergen Facts," https://www.nobelprize.org/prizes/medicine/1973/tinbergen/facts/ (accessed February 14, 2023).

15 Börje Cronholm, "Award ceremony speech," Nobel Archive, https://www.nobelprize.org/prizes/medicine/1973/ceremony-speech/ (accessed March 31, 2023).

16 Angetter, *Die österreichischen Medizinnobelpreisträger*, 68.

17 Angetter, *Die österreichischen Medizinnobelpreisträger*, 68.

18 See media reports: Föger and Taschwer, *Die andere Seites des Spiegels*, 21–30.

19 Tania Munz, *The Dancing Bees. Karl von Frisch and the Discovery of the Honeybee Language* (Chicago: The University of Chicago Press, 2016), 97–123.

20 Angetter, *Die österreichischen Medizinnobelpreisträger*, 64–66.

21 Klaus Taschwer, "Die verlorene Ehre des Konrad Lorenz," *Der Standard*, December 12, 2015, https://www.derstandard.at/story/2000027787429/die-verlorene-ehre-des-konrad-lorenz. (accessed December 1, 2023).

22 Föger and Taschwer, *Die andere Seites des Spiegels*, 19.

23 Kurt-Jürgen Voigt, "Nobelpreis für Konrad Lorenz. Gans groß gefeiert," *Spiegel*, October 4, 2010, https://www.spiegel.de/geschichte/nobelpreis-fuer-konrad-lorenz-gans-gross-gefeiert-a-948747.html, (accessed December 1, 2023)

24 Leopold Lukschanderl, "Nobelpreis für zwei Österreicher," *Österreichische Apotheker-Zeitung* 27 (1973), 49, 987.

25 Leopold Lukschanderl, "Mit K. Lorenz u. K. v. Frisch erhöht sich die Zahl der ausgezeichneten Alpenrepublikaner auf 15," *Informationsdienst für Bildungspolitik und Forschung*, no. 194 (1973), 2.

26 Lukschanderl, "Mit K. Lorenz u. K. v. Frisch," 2.

27 Föger and Taschwer, *Die andere Seites des Spiegels*, 19–20.

28 Voigt, "Nobelpreis für Konrad Lorenz."

29 Incidentally, Tinbergen is the second person in his family to be awarded a Nobel Prize. In 1969, his brother, the mathematician and economist Jan, who was four years older, received the Alfred Nobel Memorial Prize in Economic Sciences together with the Norwegian economist Ragnar Anton Kittil Frisch. "Liste der Träger des Alfred-Nobel-Gedächtnispreises für Wirtschaftswissenschaften," Wikipedia https://de.wikipedia.org/wiki/Liste_der_Tr%C3%A4ger_des_Alfred-Nobel-Ged%C3%A4chtnispreises_f%C3%BCr_Wirtschaftswissenschaften (accessed April 4, 2023).

30 Voigt, "Nobelpreis für Konrad Lorenz."

31 In 1953 Konrad Lorenz was nominated by Albrecht Peiper, professor of pediatrics in Leipzig. "Nomination for Nobel Prize in Physiology or Medicine Year: 1953," Nomination Archive, https://www.nobelprize.org/nomination/archive/show.php?id=9545 (accessed July 23, 2023).

32 Föger and Taschwer, *Die andere Seites des Spiegels*, 20.

33 Föger and Taschwer, *Die andere Seites des Spiegels*, 21, footnote 6.

34 Kruuk, *Niko's Nature*, 3.

35 Schulze-Hagen, "Tinbergen," 7.

36 Schulze-Hagen, "Tinbergen," 9.

37 Schulze-Hagen, "Tinbergen," 9.

38 Schulze-Hagen, "Tinbergen," 9.

39 Schulze-Hagen, "Tinbergen," 9.

40 Schulze-Hagen, "Tinbergen," 9.

41 Celli, *Konrad Lorenz*, 38.

42 Konrad Lorenz was co-editor of the journal. The publication was followed internationally with great interest. In a letter to the Dutch biologist and ethologist Johan Bierens de Haan, Tinbergen said about the first issue: "Of course I also saw the first issue of the journal for animal psychology. I think it looks promising; but the circular was written in such a wild Nazi frenzy that I didn't want to concern myself with it. But Tinbergen also felt that the magazine was better than nothing and subscribed to the journal for his institute." Föger and Taschwer, *Die andere Seites des Spiegels*, 70. After all, well-known scientists of the time published in this first edition, including Karl von Frisch, Konrad Lorenz, Otto Heinroth, Otto Koehler, Jakob Johann von Uexküll, Friedrich Alverdes, Hans Volkelt, Werner Fischel, Bastian Schmid and Friedrich Schwangart, *Zeitschrift für Tierpsychologie* 1 (1), (January-December 1937), 1–96, https://onlinelibrary.wiley.com/toc/14390310a/1937/1/1 (accessed December 1, 2023).

43 Konrad Lorenz and Niko Tinbergen, "Taxis und Instinkthandlung in der Eirollbewegung der Graugans," *Zeitschrift für Tierpsychologie* 2 (1), 1–29, http://klha.at/papers/1938-Eirollbewegung.pdf (accessed December 1, 2023). – The first footnote underlines again the different working methods of the two authors: The occasion for this work, which was carried out with the support of the foundation "Het Donderfonds," came from discussions held in Leiden and later in Altenberg on the concept of instinctual action established by Lorenz (1937). Although the share of the individual can hardly be limited, it should be mentioned that the theoretical explanations come to the greater part from Lorenz, the invention and execution of the experiments, however, represents mainly Tinbergen's share.

44 Hubert J. Gieß, "Verhaltensforschung: Theorie ohne Wert?" *Psychologie heute* (Juli 1993), 9–10.

45 Hanna-Maria Zippelius, *Die vermessene Theorie. Eine kritische Auseinandersetzung mit der Instinkttheorie von Konrad Lorenz und verhaltenskundlicher Forschungspraxis* (Braunschweig/ Wiesbaden: Vieweg, 1992).

46 Hubert J. Gieß, "Verhaltensforschung," 9–10.

47 In this series of experiments, Lorenz and Tinbergen observed the reactions to silhouettes of flying large birds. Lorenz described this in 1939 as follows: "Tinbergen and I [...] worked on almost all [...] available young birds [ducks, geese, turkeys] with dummy birds of prey, which were moved on a cable car stretched between two tall trees." – Schulze-Hagen, "Tinbergen," 10. If such a cardboard disc was pulled forward, it looked like a bird of prey (short neck and long tail) and triggered alarm reactions. When the cardboard disc was pulled backwards, it resembled the flight pattern of geese (long neck, short tail), and the alarm was absent. Lorenz and Tinbergen were inspired to conduct such studies by Heinroth's observations at the Berlin Zoo. – Nikolaas Tinbergen, *The Study of Instinct* (Oxford: Clarendon Press, 1951), 30–31.

48 Schulze-Hagen, "Tinbergen," 10.

49 Konrad Lorenz, "My family and other animals," in *Studying animal behavior. Autobiographies of the Founders*, edited by Donald A. Dewsbury (Chicago/London: Chicago University Press, 1985), 269.

50 Celli, *Konrad Lorenz*, 32.

51 Schulze-Hagen, “Tinbergen,” 20.

52 Gerhard Hirschfeld, “Die Universität Leiden unter dem Nationalsozialismus,” *Geschichte und Gesellschaft* 23 (1997), 560–591.

53 Kruuk, *Niko's Nature*, 115.

54 Schulze-Hagen, “Tinbergen,” 19, footnote 19.

55 Schulze-Hagen, “Tinbergen,” 13–14.

56 Schulze-Hagen, “Tinbergen,” 12.

57 Konrad Lorenz called the camp a concentration camp, but in reality, it was an internment camp.

58 Schulze-Hagen, “Tinbergen,” 12.

59 Föger and Taschwer, *Die andere Seites des Spiegels*, 21–22.

60 Benedikt Föger, “Von ‘Lebenstuchtigen’ und ‘Rassenpflege’: Über zwei Nobelpreisträger aus Österreich,” *Die Presse, Spectrum*, (December 5, 1998), XIII.

61 Celli, *Konrad Lorenz*, 57.

62 Föger and Taschwer, *Die andere Seites des Spiegels*, 33–34.

63 See for example: Theodora J. Kalikow, “Konrad Lorenz’ ‘Brown Past’: A Reply to Alec Nisbett,” Journal of the History of Behavioural Science 14 (1973), 173–179. – Ute Deichmann, *Biologen unter Hitler. Porträt einer Wissenschaft im NS-Staat* (Frankfurt am Main: Fischer TB, 1995), 295–297.

64 Theodora J. Kalikow, “Konrad Lorenz's ethological theory: Explanation and ideology 1938–1943,” *Journal of the History of Biology* 16, no. 1 (1983), 52–53.

65 Kurt Kotrschal, “Mit Federn, Haut und Haar: Kein ‘Vater der Graugänse’. Konrad Lorenz bereitete den Boden für die Synthese von Ethologie und Psychologie.” *Die Presse*, February 24, 2009, https://www.diepresse.com/455178/mit-federn-haut-und-haar-kein-bdquovater-der-graugaenseldquo (accessed March 31, 2023). See also: Kurt Kotrschal, “NS-Eugenik und Konrad Lorenz. Nicht ‘bekennend’,” *Die Presse, Spectrum*, (December 12, 1998), XV.

66 “Beschluss vom 15.12.2015 des Senats der Paris Lodron Universität Salzburg im Einvernehmen mit dem Rektorat über die Aberkennung von Ehrungen,” (Resolution of December 15, 2015 of the Senate of the Paris Lodron University of Salzburg in agreement with the Rectorate on the withdrawal of honors, Translation Angetter), Paris Lodon Universität Salzburg, https://b.ds.at/2015/12/18/HonorumSenat01_1.pdf (accessed February 15, 2023).

67 Kotrschal, “ NS-Eugenik,” XV.

68 “Konrad Lorenz Biografie,” Wien Geschichte Wiki, https://www.geschichtewiki.wien.gv.at/Konrad_Lorenz (accessed April 4, 2023).

69 Konrad Lorenz, *Die acht Todsünden der zivilisierten Menschheit.* (Piper; München, Zürich 1973), 64.

70 Lorenz, *Die acht Todsünden*, 57.

71 Föger and Taschwer, *Die andere Seites des Spiegels*, 25, footnote 11.

72 Schulze-Hagen, “Tinbergen,” 27.

73 Konrad Lorenz, *Gespräche mit Richard I. Evans, ein Briefwechsel mit Donald Campell und vier Essays*, edited Richard I. Evans (Frankfur/M, Berlin: Ullstein Verlag, 1977), 10.

74 Föger and Taschwer, *Die andere Seites des Spiegels*, 206, footnote 22.

75 Föger and Taschwer, *Die andere Seites des Spiegels*, 30 footnote 24.

76 Föger and Taschwer, *Die andere Seites des Spiegels*, 33.

77 “Wir werden von Steinzeitmenschen regiert,” Spiegel-Gespräch mit dem Verhaltensforscher Konrad Lorenz über die Zukunft der Menschheit, *Der Spiegel* 45 (1988), https://www.spiegel.de/kultur/wir-werden-von-steinzeitmenschen-regiert-spiegel-gespraech-mit-a-1f0255cd-0002-0001-0000-000013530460 (accessedMarch 31, 2023).

78 Nobel Archive, “Award ceremony speech,” https://www.nobelprize.org/prizes/medicine/1973/ceremony-speech/ (accessed April 4, 2023).

79 Munz, *The Dancing Bees*, 4.

80 The decree on the installation of the Rassenpolitisches Amt stated: "Apart from the standardization and monitoring of training and propaganda in the relevant areas, the Office's scope of duties also includes all factual, population and racial policy issues, insofar as they are dealt with by the party" (Translation Angetter). IfZ-München, Akten d. Parteikanzlei, 117 04801, May 15, 1934. Quoted from: Roger Uhle, Neues Volk und reine Rasse. Walter Gross und das Rassenpolitische Amt der NSDAP (RPA) 1934–1945 (PhD diss. Rheinisch-Westfälisch Technische Hochschule Aachen, 1999), 30.

81 Taschwer, "Der Bienenforscher und das NS-Regime."

82 Taschwer, "Der Bienenforscher und das NS-Regime."

83 Wolfgang Neugebauer, "Zur Problematik der NS-Vergangenheit Österreichs," Dokumentationsarchiv des österreichischen Widerstandes, https://www.doew.at/erforschen/projekte/arbeitsschwerpunkte/widerstand-und-verfolgung/umgang-mit-der-ns-vergangenheit/zur-problematik-der-ns-vergangenheit-oesterreichs (accessed April 4, 2023).

84 In this progress report another Austrian Nobel Prize winner for physiology or medicine also came under criticism: Julius Wagner-Jauregg (1857–1940). The 1927 Nobel Prize winner for discovery of the therapeutic significance of malaria vaccination in the treatment of "progressive paralysis" had been a member of the anti-Semitic Greater German Party since 1937 and had sympathized with National Socialism. In addition, it was precisely his experiments on malaria therapy that paved the way for later inhuman malaria experiments at the clinic of psychiatrist Hans Hoff (1897–1969) in Vienna. It is true that Wagner-Jauregg was not involved in medical or other Nazi crimes; after all, he was already over 80 years old at the time of the "Anschluss," and his application for membership in the NSDAP on April 21, 1940, was postponed due to the Jewish origin of his first wife. But Wagner-Jauregg vehemently advocated continuing the tradition of the Viennese medical school in a Jew-free form. When the Society of Physicians was dissolved in 1938 and re-established as Vienna Medical Society, Wagner-Jauregg was the first to apply for admission to the new society. – Angetter, *Die österreichischen Medizinnobelpreisträger*, 42.

85 Beschluss vom 15.12.2015 des Senats der Paris Lodron Universität Salzburg.

86 "ANFRAGE des Abgeordneten Dr. Andreas F. Karlsböck und weiterer Abgeordneter an den Bundesminister für Wissenschaft, Forschung und Wirtschaft betreffend Grundlagen der Verleihung und Aberkennung von Ehrendoktoraten durch die Universität Salzburg," Parlamentarische Anfrage (January 14, 2016), 7610/JXXV.GP, https://www.parlament.gv.at/dokument/XXV/J/7610/fnameorig_497232.html (accessed July 21, 2023).

87 Beschluss vom 15.12.2015 des Senats der Paris Lodron Universität Salzburg [Resolution of December 15, 2015 of the Senate of the Paris Lodron University of Salzburg].

## Bibliography

Angetter, Daniela. "Die österreichischen Medizinnobelpreisträger." In *Österreichisches Biographisches Lexikon, Schriftenreihe 8.* Wien: Verlag der österreichischen Akademie der Wissenschaften, 2003, 58–70.

Celli, Giorgio. *Konrad Lorenz. Begründer der Ethologie.* Heidelberg: Spektrum der Wissenschaft, 2001.

Cronholm, Börje. "Award ceremony speech." https://www.nobelprize.org/prizes/medicine/1973/ceremony-speech/ (accessed March 31, 2023).

Deichmann, Ute. *Biologen unter Hitler. Porträt einer Wissenschaft im NS-Staat.* Frankfurt am Main: Fischer TB, 1995.

Evans, Richard I., ed. *Konrad Lorenz. Gespräche mit Richard I. Evans, ein Briefwechsel mit Donald Campell und vier Essays.* Frankfurt/M., Berlin, Wien: Ullstein Verlag, 1977.

Föger, Benedikt and Klaus Taschwer. *Die andere Seites des Spiegels. Konrad Lorenz und der Nationalsozialismus.* Wien:Czernin Verlag, 2001.

Föger, Benedikt. "Von 'Lebenstuchtigen' und 'Rassenpflege': Über zwei Nobelpreisträger aus Österreich." *Die Presse,* Spectrum (December 5, 1998).

Geschichte Wiki, "Konrad Lorenz Biografie." https://www.geschichtewiki.wien.gv.at/Konrad_Lorenz (accessed April 4, 2023).

Gieß, Hubert J. "Verhaltensforschung: Theorie ohne Wert?" *Psychologie heute* (Juli 1993): 9–10.

Hirschfeld, Gerhard. "Die Universität Leiden unter dem Nationalsozialismus," *Geschichte und Gesellschaft* 23 (1997): 560–591.

Kalikow, Theodora J. "Konrad Lorenz's ethological theory: Explanation and ideology 1938–1943." *Journal of the History of Biology* 16, no. 1 (1983): 52–53.

Kalikow, Theodora J. "Konrad Lorenz' 'Brown Past': A Reply to Alec Nisbett." *Journal of the History of Behavioural Science* 14 (1973): 173–179.

Karlsböck, Andreas F. "Anfrage des Abgeordneten Dr. Andreas F. Karlsböck und weiterer Abgeordneter an den Bundesminister für Wissenschaft, Forschung und Wirtschaft betreffend Grundlagen der Verleihung und Aberkennung von Ehrendoktoraten durch die Universität Salzburg, " Parliamentary question received on January 14, 2016, 7610/JXXV.GP. https://www.parlament.gv.at/dokument/XXV/J/7610/fnameorig_497232.html.

Kotrschal, Kurt. "NS-Eugenik und Konrad Lorenz. Nicht 'bekennend'," *Spectrum* (December 12, 1998): XV.

Kotrschal, Kurt. "Mit Federn, Haut und Haar: Kein 'Vater der Graugänse'. Konrad Lorenz bereitete den Boden für die Synthese von Ethologie und Psychologie." *Die Presse* (Feburary 23, 2009). https://www.diepresse.com/455178/mit-federn-haut-und-haar-kein-bdquovater-der-graugaenseldquo (accessed March 31, 2023).

Kruuk, Hans. *Niko's Nature. The Life of Niko Tinbergen and his Science of Animal Behaviour.* Oxford: Oxford University Press, 2003.

Lorenz, Konrad and Niko Tinbergen. "Taxis und Instinkthandlung in der Eirollbewegung der Graugans," *Zeitschrift für Tierpsychologie* 2 (1), 1938: 1–29, http://klha.at/papers/1938-Eirollbewegung.pdf.

Lorenz, Konrad. "Wir werden von Steinzeitmenschen regiert," Spiegel-Gespräch mit dem Verhaltensforscher Konrad Lorenz über die Zukunft der Menschheit, Der Spiegel 45 (1988), https://www.spiegel.de/kultur/wir-werden-von-steinzeitmenschen-regiert-spiegel-gespraech-mit-a-1f0255cd-0002-0001-0000-000013530460 (accessed March 31, 2023).

Lorenz, Konrad. *Die acht Todsünden der zivilisierten Menschheit.* München, Zürich: Piper, 1973.

Lorenz, Konrad. "My family and other animals." In *Studying animal behavior. Autobiographies of the Founders,* edited by Donald A. Dewsbury, 269. Chicago, London: Chicago University Press, 1985.

Lukschanderl, Leopold. "Nobelpreis für zwei Österreicher." *Österreichische Apotheker-Zeitung* 27 (1973): 987.

Lukschanderl, Leopold. "Mit K. Lorenz u. K. v. Frisch erhöht sich die Zahl der ausgezeichneten Alpenrepublikaner auf 15'." *Informationsdienst für Bildungspolitik und Forschung* no. 194 (1973): 2.

Max-Planck-Gesellschaft, "Nobelpreis 2022 für Svante Pääbo." https://www.mpg.de/19316167/nobelpreis-fuer-svante-paeaebo (accessed March 31, 2023).

Munz, Tania. *The Dancing Bees: Karl von Frisch and the Discovery of the Honeybee Language.* Chicago: The University of Chicago Press, 2016.

Neugebauer, Wolfgang. "Zur Problematik der NS-Vergangenheit Österreichs," *Dokumentationsarchiv des österreichischen Widerstandes.* https://www.doew.at/erforschen/projekte/arbeitsschwerpunkte/widerstand-und-verfolgung/umgang-mit-der-ns-ve rgangenheit/zur-problematik-der-ns-vergangenheit-oesterreichs (accessed April 4, 2023).

Nobel Archive. "Award ceremony speech." https://www.nobelprize.org/prizes/medicine/1973/ceremony-speech/ (accessed April 4, 2023).

Nomination Archive. "Karl von Frisch Facts." https://www.nobelprize.org/prizes/medicine/1973/frisch/facts/ (accessed February 14, 2023).

Nomination Archive. "Nikolaas Tinbergen Facts." https://www.nobelprize.org/prizes/medicine/1973/tinbergen/facts/ (accessed February 14, 2023).

Nomination for Nobel Prize in Physiology or Medicine Year: 1953. Nomination Archive. https://www.nobelprize.org/nomination/archive/show.php?id=9545.

Paris Lodron Universität Salzburg, Beschluss vom 15.12.2015 des Senats im Einvernehmen mit dem Rektorat über die Aberkennung von Ehrungen [Resolution of December 15, 2015 of the Senate of the Paris Lodron University of Salzburg in agreement with the rectorate on the withdrawal of honours]. Paris Lodron Universität Salzburg, https://b.ds.at/2015/12/18/HonorumSenat01_1.pdf (accessed February 15, 2023).

Schulze-Hagen, Karl. "Tinbergen und deutsche Ornithologen: Eine wechselseitige Inspiration." *Vogelwarte* 59 (2001): 7.

Taschwer, Klaus. "Der Bienenforscher und das NS-Regime." *Der Standard* (December 2014). https://www.derstandard.at/story/2000009906567/der-bienenforscher-und-das-ns-regime.

Taschwer, Klaus. "Die verlorene Ehre des Konrad Lorenz." *Der Standard* (December 12, 2015) https://www.derstandard.at/story/2000027787429/die-verlorene-ehre-des-konrad-lorenz.

Tinbergen, Nikolaas. "Watching and wondering." In *Studying animal behavior. Autobiographies of the Founders*, edited by Donald A. Dewsbury, 431–463. Chicago, London: Chicago University Press, 1985.

Tinbergen, Nikolaas. *The Study of Instinct.* Oxford: Clarendon Press, 1951.

Uhle, Roger. Neues Volk und reine Rasse. Walter Gross und das Rassenpolitische Amt der NSDAP (RPA) 1934–1945. PhD diss., Rheinisch-Westfälisch Technische Hochschule Aachen, 1999.

Voigt, Kurt-Jürgen. "Nobelpreis für Konrad Lorenz. Gans groß gefeiert." *Spiegel* (October 4, 2010), https://www.spiegel.de/geschichte/nobelpreis-fuer-konrad-lorenz-gans-gross-gefeiert-a-948747.html. von Frisch, Karl. *Erinnerungen eines Biologen*, 3rd edition. Berlin, Heidelberg, New York: Springer, 1973.

Zippelius, Hanna-Maria. *Die vermessene Theorie. Eine kritische Auseinandersetzung mit der Instinkttheorie von Konrad Lorenz und verhaltenskundlicher Forschungspraxis.* Braunschweig, Wiesbaden: Vieweg, 1992.

CHAPTER TEN

# What Does it Take for an Anatomist to Get a Nobel Prize? An Analysis of the Nobel Prize Nominations for Wolfgang Bargmann, Albert von Kölliker and Hans Spemann

*Nils Hansson, Giacomo Padrini, Andreas Winkelmann, Mathias Schütz*

**Abstract**

Only one Nobel Prize has been awarded for anatomical work so far: Camillo Golgi and Santiago Ramón y Cayal shared the 1906 prize "for their work on the structure of the nervous system." Why so few? This study sheds light on anatomists who were nominated for the Nobel Prize but did not receive it and discusses why well-known scholars were turned down by the Nobel Committee in Stockholm. An analysis of Nobel archive files and Nobel Prize nominations will help us to add new perspectives on the reputation of individual scientists and to reconstruct important scientific trends during the 20th century, as exemplified by the field of anatomy.

**Keywords:** Anatomical Research, Anatomical Society, Anatomy and Science, Scientific Recognition, History of Anatomy

## Introduction: The *Nobel population* in anatomy

On the face of it, only one Nobel Prize has been awarded for anatomical work so far: Camillo Golgi and Santiago Ramón y Cajal shared the 1906 prize "for their work on the structure of the nervous system." Why so few? This study sheds light on anatomists who were nominated for the Nobel Prize but did not receive it and discusses why well-known scholars were turned down by the Nobel Committee in Stockholm. An analysis of Nobel archive files and Nobel Prize nominations will help us to add new perspectives on the reputation of individual scientists and to reconstruct important scientific trends during the 20th century, as exemplified by the field of anatomy.[1]

The Nobel laureate selection process has scarcely changed since its inception at the turn of the 20th century.[2] The process has three main steps. First, scientists from different countries are invited by the Nobel Committee to submit proposals (self-nominations are not allowed). Some scholars are entitled to nominate every year, like previous Nobel laureates and members of the Nobel Assembly. The Assembly consists of fifty voting members, all of them professors in medicine at the Karolinska Institute. The Assembly also selects its working body, the Nobel Committee with around five members. After the nomination deadline, the Committee invites experts to evaluate the nominees. During the first half of the 20th century, most evaluations were written by members of the Committee or the Assembly. These evaluations were then discussed within the Committee, which submitted its recommendation(s) to the Assembly. Finally, the Assembly chose the laureate(s) by a majority vote.

## Methods

The Nobel nomination database (nobelprize.org) is currently accessible online for the period from 1901 to 1953 and includes 5950 nominations in physiology or medicine. In a first approach, we extracted all relevant nominees by using the keywords "anatomy"/"anatomical," "histology"/"histological," "morphology"/"morphological," and "embryology"/"embryological," then by focusing on German scholars. The results were already indicative of the larger challenge of investigating the representation of anatomy in the history of the Nobel Prize, namely, that a majority of scientists who received nominations citing histological or anatomical work as its main reason conducted research in other fields, especially pathology (e.g. Ludwig Aschoff and Félix Marchand) and physiology (e.g. Eduard Hensen and Victor Strassburger). Another example is the German zoologist and anatomist Theodor Boveri, known for ground-breaking work on cellular processes central to the etiology of cancer, who was nominated in 1909, but for his general work on cytology and the chemistry of fertilization. Additionally, differences in the use of the terms "anatomy" and "pathology" in the languages of the nominators must be considered. Whilst in German, anatomy and pathology are usually separate subjects, those may be referred to as "physiological anatomy" and "pathological anatomy" in English. This makes it a delicate task to define anatomy as a category in a Nobel Prize context.

All these considerations point to a fundamental question challenging the Nobel history of anatomy: who is, in fact, an anatomist? The disciplinary disentanglement of human anatomy, physiology, and pathology in the course of the 19th century cannot hide a common root: A common, "anatomical idea" in the words of Rudolf Virchow, of relating functional, pathological, and developmental aspects of the organism to its form and its localities.[3] Furthermore, as Lynn Nyhart has shown, anatomy was deeply embedded in the conception of scientific concepts and techniques – here the rise of Darwinism and experimental developmental biology – which transcended the narrow jurisdiction of anatomists within the German medical schools.[4] Both aspects point to an increasing vagueness of the term anatomy and its disciplinary boundaries which make it difficult to definitely distinguish anatomists considered for the Nobel Prize, a vagueness that is also reflected in career paths which fluctuated between medicine and biology, between morphological, histological, physiological, embryological, and cytological research without attempting or being able to draw clear-cut lines between scientific fields and methods. To circumvent this difficulty, we have also applied membership of the Anatomische Gesellschaft as a criterion that demonstrates who was seen as an anatomist by himself and his peers at the time. The Anatomische Gesellschaft, still in existence, was founded in 1886 as a German-speaking society with an international aspiration and therefore avoided the term "German" in its name. From its very beginning, it defined itself with a rather broad "objective demarcation of the field," explicitly encompassing in its first bylaws "the anatomical sciences in their entire range (including histology, developmental history, comparative anatomy, etc.)."[5]

As this first approach showed that defining the exact scope of "anatomical research" would be too difficult to allow an all-encompassing analysis of the field, we decided for an in-depth analysis of three exemplary candidates. For these, we collected documents like nomination letters and jury reports – that are currently accessible in the archive of the Nobel Committee in Sweden (on site accessibility is for documents up to 1973).

To make the historical analysis of such a disciplinary range of candidates more concise and connect it to the history of anatomy as a scientific field, we highlight three German nominees: Albert von Kölliker, the first scholar nominated for anatomical work, Nobel laureate Hans Spemann, and a leading anatomist of post-war West-Germany, Wolfgang Bargmann. To put those nominations and the evaluation of their research by the Nobel committee into a broader context, we have paired each of those scholars with respective international colleagues: the 1906 laureates Camillo Golgi and Santiago Ramón y Cajal, the German refugee scholar Ernst Scharrer, and Yale zoologist Ross Granville Harisson. These examples shed light on the shifting status of anatomy as a particular field of research in the eyes of the (biomedical) scientific community in the first half of the twentieth century, as much as it highlights what kind of research was deemed innovative and laudable from the perspective of the Nobel committee.

## Results

Four countries stand out in terms of nominees in the first half of the 20th century: France, Germany, the United Kingdom and the United States. Already the range of nominees from Germany sheds light on the range of scientific interests and methods developed by contemporary anatomists, ranging from Carl Gegenbaur's comparative anatomy via Wilhelm Roux's developmental mechanics[6] to neuroanatomist Cécile Vogt, the first woman nominated for the Nobel Prize in physiology or medicine in 1922.

## Albert von Kölliker

Born Rudolf Albert Kölliker in Switzerland in 1817, Kölliker attended medical school in Zürich, Bonn, and Berlin. Among his teachers was Jakob Henle, the Bavarian anatomist and histologist.[7] They met in Zürich where Henle held the chair in anatomy from 1840 to 1884. Kölliker became Henle's research assistant. Henle, known for his description of the renal tubular system, returned to Germany in 1844. The Zürich chair of anatomy was subsequently split and Kölliker was appointed as extraordinary professor of physiology and comparative anatomy only two years after he had completed his MD. As soon as in 1847, Kölliker became full professor in physiology and comparative anatomy at Würzburg university. In 1856, he was also appointed chair of comparative and topographical anatomy at the same university. In subsequent years he gradually reduced his field of activity, eventually limiting his work to anatomy and histology, subjects in which he continued to conduct research even after his retirement in 1897, the same year he was raised to the ranks of aristocracy by prince Luitpold, changing name to Albert *von* Kölliker.[8] In 1905, von Kölliker died in Würzburg.

Von Kölliker is remembered as one of the founding fathers of modern histology, for his findings in embryological organogenesis as well as for having studied various tissues of the human body. He was also among the founders of the Anatomische Gesellschaft in 1886, served as the society's first president, and was appointed as "permanent honorary chairman" in 1894.[9] From 1901 to his death in 1905, von Kölliker was nominated for the Nobel Prize in four nomination letters. Among his nominators were fellow anatomists like Viktor von Ebner and Karolinska Institute member Gustav Retzius. All his proponents emphasized his 60-year career in anatomy and general work on histology, whilst one scholar mentioned his contributions to embryology. However, von Kölliker was deemed too old by the committee, and it is well-known that the Nobel Prize is not usually awarded for lifetime achievements – the work he was most renowned for, his *Handbuch der Gewebelehre*, had first been published in 1852.

Still, Albert von Kölliker remains interesting in a Nobel Prize context, especially when considering his role as a nominator. In 1901 von Kölliker proposed Camillo Golgi for his "extensive work on the central nervous system" and his method of staining cells and in 1905 as well as 1906 he tried again, this time also including Santiago Ramón y Cajal, therefore nominating the two scholars who succeeded in getting the 1906 Nobel Prize as the only two anatomists in history (Nobel yearbook 1901). Both Golgi and Ramón y Cajal had joined von Kölliker early on in the ranks of the Anatomische Gesellschaft: The Italian Golgi, professor first of histology and then of pathology at the University of Pavia, became a member in 1889, and the Spaniard Ramón y Cajal, who held professorships of anatomy, histology and pathology at the Universities of Valencia, Barcelona and Madrid, joined the society first in 1893, had his membership cancelled in 1901 for continuous non-payment of his fees, and re-joined the Anatomische Gesellschaft in 1907, the year after receiving the Nobel Prize together with Golgi, now acquiring the status of a lifelong member. Their findings were disputed at the time of their discovery and von Kölliker proved to be one of their fiercest advocates.[10] His relationship to Golgi can be characterised as a close friendship.[11] In his 1905 nomination letter, he referred to Golgi and Ramón y Cajal as the "most important promoters of microscopy" (Nobel yearbook 1905). Whilst drawing a very positive image of the nominee is not unusual for a Nobel nomination, von Kölliker surprised the readers by then continuing with a lengthy tirade against other anatomists claiming that Albrecht Bethe had fundamentally erred in his assessment of nervous tissue and claiming that Franz Nissl had not based his publications on scientific observations. Having said that, von Kölliker gave little support of his criticism, merely stating that Bethe's claims about neuronal regeneration had been proven wrong by Italian researchers like Aldo Perroncito. Von Kölliker concluded his letter by referring to his own most recent publication of which he included some copies for the Nobel Committee. When the decision of the committee to actually award the 1906 Nobel Prize to Golgi and Ramón y Cajal was published, von Kölliker had died. Von Kölliker helped lay the foundation for Golgi and Ramón y Cajal receiving the Nobel Prize, not only through his support for their scientific findings and their nomination, but by shaping the field of histology, its concepts and techniques, in which they were to succeed. Although von Kölliker's own scientific legacy was much broader, the specific, technical and conceptual contribution of Golgi and Ramón y Cajal to the understanding of the nervous system proved to be more fitting to be eligible for a Nobel Prize.

### Wolfgang Bargmann – Nobel Prize candidate in the 1950s and 1960s

While such a specific contribution to a larger field of research was deemed laudable at the beginning of the century, the constellation had been reversed fifty years later as the example of Wolfgang Bargmann shows. Born in 1906, Bargmann's interest for natural sciences and especially for tissues and structures of organs emerged when he was still in school. Bargmann studied zoology and medicine in various European cities, graduating in medicine from Frankfurt University. In

1931, he received his MD degree with a thesis on the comparative anatomy of the fine structure of human and mammalian kidneys. After conducting research in Frankfurt, Bargmann joined Freiburg University in 1934 where he became assistant to Wilhelm von Möllendorff, who also taught the later Nobel Laureate Hans Adolf Krebs.[12] Von Möllendorff, who had disapproved of the National Socialist government's policies increasingly threatening his Jewish colleagues (among them Krebs) emigrated to Switzerland in 1935 where he had been offered the chair in anatomy at Zürich University. Wolfgang Bargmann followed him and stayed at his institute until 1938 when he joined Max Clara's institute in Leipzig.[13] In Clara, Bargmann had found a strong supporter. Finally, Bargmann became chairman of the anatomical department of Königsberg University. Having to leave Königsberg at the end of World War II, Bargmann, after a short stop in Göttingen, became head of the Anatomical Department at Kiel University where he stayed until his retirement in 1974.[14] He had been a member of the Anatomische Gesellschaft since 1934.[15] In 2013 the Gesellschaft renamed its Bargmann prize because of Bargmann's involvement in the use of the bodies of executed Nazi victims for his research.

The first Nobel Prize nomination of Wolfgang Bargmann was submitted in 1955 (Nobel yearbook 1955). Among his nominators was the Kiel dermatologist Albin Proppe, who also put Feodor Lynen's name forward for his work on lipid metabolism. Among Bargmann's work, Proppe judged the morphological description of hypophysal hormone production as prizeworthy. He did not praise the scholars he nominated in the hyperbolic way many other nominators did, but ended his letter with the statement "I presume that the achievements of both researchers are acknowledged by proven experts, hence I request to refrain from enclosing the pertinent literature." It is highly probable that Bargmann and Proppe knew each other personally as they were both professors at Kiel University in the 1950s.[16]

Bargmann received more nominations in the following years, from researchers from all over Europe, like Greek physician Th. Sklavounos. Nominating Bargmann for the 1963 Nobel Prize, Sklavounos complained that anatomy had been neglected by the Nobel Committee, which, in his view, concentrated too much on biochemistry, immunology, and physiology, not acknowledging anatomy and histology with their new methods like the electron microscopy. After this introduction of roughly a page, he went on to name Bargmann as his candidate and added a four-page list of Bargmann's publications including both scientific work and textbooks. Whilst he did comment on some points, mentioning discoveries about secretory cells of the kidney as well as lauding Bargmann for his teaching skills (mentioning his textbook *Histologie und mikroskopische Anatomie des Menschen*) and for editing the journal "Zeitschrift für Zellforschung", Sklavounos did not pinpoint a single discovery as prizeworthy. Instead, he concluded that Bargmann should be awarded the Nobel Prize for being a "brilliant research personality." He praised Bargmann's merits in the service of anatomy further by stating that he made the anatomical department of Kiel University a venue for international scholars.

Once a candidate is deemed interesting by the Nobel committee, the next step in the selection process are written reports by experts. In 1955, the Stockholm anatomist Gösta Häggqvist wrote a six-page statement for the Nobel Committee about Bargmann.[17] After a historical review of other pioneers regarding hormone production (only mentioning the surnames Oliver, Schaefer, Dale, Kamm, and Speidel), he focused on Bargmann's publications from 1940 onward. Häggqvist was quite enthusiastic about Bargmann's research, in particular his use of the hematoxylin-phloxine-method, but emphasized that Ernst Scharrer would be the stronger candidate. Scharrer, who studied both zoology and medicine in Munich, receiving his PhD in 1928 and his MD in 1933, subsequently worked in both fields at Yale University and in Vienna, Munich and Frankfurt. In 1937, Ernst Scharrer was forced to emigrate to the United States together with his wife and scientific collaborator Berta, where he worked as an anatomist at different medical schools: briefly, after his arrival, at the University of Chicago, from 1938 to 1940 at the Rockefeller Institute in New York City, from 1940 to 1945 at Case Western Reserve University in Cleveland and from 1946 to 1955 at the University of Colorado Medical School in Denver, until he returned to New York City in 1955 where he and his wife became professors at the newly founded Albert Einstein College of Medicine. He joined the Anatomische Gesellschaft only after its post-war reconstitution, in 1951 (Stieve and Watzka 1951, 215). Already in Frankfurt in the 1930s, Ernst and Berta Scharrer had met and worked with Bargmann, sparking the latter's interest in the concept of neurosecretion, which all of them would investigate and publish on in the following years, even collaboratively.[18] Scharrer was, in the words of Bargmann, "zoologist and anatomist in one person" (Bargmann 1966, 123), representing an interconnectedness of both fields that was felt and lived by many of their peers – and that, incidentally, contributed to the perception of anatomy as a past science since many innovations anatomists produced would step out of the anatomical realm and form their own scientific school, infrastructure and, eventually, disciplinary boundaries. Here, Scharrer's and Bargmann's shared research into neurosecretion is a case in point, as the latter's considerations for the Nobel Prize would highlight. Häggqvist, in his 1955 statement on Bargmann, stressed the specific fact that Scharrer had been invited to give the keynote lecture at the 1953 Anatomische Gesellschaft meeting, whereas Bargmann "only" provided a comment ("Koreferat"), obviously assessing their respective scientific stature and worthiness of being considered for the Nobel Prize. Häggqvist also added that Bargmann had asked Scharrer to write the chapter on neurosecretion in the *Handbuch der mikroskopischen Anatomie*. In 1963, Bargmann was evaluated again by the Nobel Committee, this time by the physiologist Ulf Svante von Euler. Euler did not deem Bargmann prizeworthy, and suggested that Scharrer (the same year, Scharrer published the widespread textbook *Neuroendocrinology*) and Ernest Basil Verney were stronger candidates in this field. While Bargmann had specified the process of neurosecretion through morphological and histological techniques, it was now the foundation and demarcation of neuroendocrinology that was increasingly considered as worthy of the Nobel Prize – although, eventually, neither Bargmann nor Ernst and Berta Scharrer would receive it.

## Hans Spemann – Nobel laureate in 1935

The zoologist Hans Spemann's investigations into the determination of embryological development not only connected him to a broader discourse on a perceived crisis and reformation of medicine, prevalent in the Weimar Republic,[19] it also connected the classical anatomical focus on ontogenesis with modern experimental techniques – a connection of interests and methods that pervades Spemann's career and is displayed in his long membership in the Anatomische Gesellschaft, which he began as a young assistant at the University of Würzburg's zoological institute in 1901 and ended for unknown reasons in 1934.[20] Spemann shared a deep connection with contemporary anatomy not only through his scientific formation, methods, and interests, but also his personal relationships. He was influenced mainly by Carl Gegenbaur, under whom he studied medicine in Heidelberg and whose comparative morphology introduced him to questions of development and descendance, but also by Wilhelm Roux, whose developmental mechanics formed an experimental counterprogramme to Gegenbaur's descriptive approach.[21] As a young assistant at the University of Würzburg's zoological institute, Speman befriended anatomist Hermann Braus, with whom he shared not only the morphological training by the "Gegenbaur school" (Nyhart 2003), but also his experimental ambitions.[22] When, under the supervision of Roux, a department for developmental mechanics was established at the newly founded Kaiser-Wilhelm-Institute for Biology in Berlin, the contenders were Braus and Spemann, the latter of whom accepted the position in 1914.[23] After the First World War, Spemann took over the professorship for zoology at the University of Freiburg, but the ties to his anatomist colleagues only intensified. He joined Roux and Braus in editing the *Archiv für mikroskopische Anatomie und Entwicklungsmechanik*, a unification of two classical anatomical and embryological journals, in which he also published his most prominent work on the organizer effect in embryological differentiation, together with his doctoral student Hilde Mangold.[24] After the death of both Braus and Roux in 1924, Spemann made sure that their legacy was sustained in several ways. Not only did he rebrand the journal as *Roux' Archiv*, and dedicate the first issue published under the new name to the memory of Hermann Braus; he also brought a subsequent generation of anatomists into the editorial board: Benno Romeis (1888–1971), professor of experimental biology at the anatomical institute of Munich University, and Walther Vogt, who had been an assistant at Braus' institute in Würzburg and was considered as Speman's successor at the Kaiser-Wilhelm-Institute for Biology, but had opted instead for a professorship for histology and embryology, also at Munich anatomy. The relationship between Spemann and his fellow anatomists was characterized by professional as much as personal appreciation, as the many mutual appraisals speak to.[25] Unsurprisingly, the ratio of anatomists among Spemann's nominators for the Nobel Prize was high: Braus had already nominated him in 1924, Vogt reiterated the nomination in 1934, shortly after another anatomist, Otto Veit of Cologne university, had done so in 1932. In 1935, the year that Spemann received the Nobel Prize for Physiology or Medicine, he had been nominated by the Swedish anatomists and histologists Erik Agduhr of Uppsala and Gösta Häggqvist of Stockholm, and by Alfred Benninghoff of Kiel University, another German anatomist who came from the Gegenbaur school and who attempted to amplify anatomy's morphological tradition with causal and functional explanations, like those developed by Spemann.[26]

Spemann was reviewed by the Nobel Committee in 1924, 1932, 1933, 1934 and 1935 (Nobel yearbook 1935). All the reviewers (Agduhr, Hesser, Häggqvist) considered him worthy of the prize. Häggqvist writes in 1935 (translation from Swedish): "In conclusion, I only want to emphasize that, the last few years, progress in developmental mechanics has confirmed the extraordinary importance of Spemann's work. I consider him highly deserving of the Nobel Prize." This time the Nobel Committee agreed and Spemann received the Nobel Prize in physiology or medicine in 1935 "for his discovery of the organizer effect in embryonic development."

Considering the ambivalence of anatomical affiliation that Spemann represents, it is striking that the only scientist he himself nominated for the Nobel Prizes was Ross Granville Harrison in 1937. Like Spemann, Harrison is another example of the contemporary fluctuations between scientific disciplines, fields of research, and methods which pointed at once towards and away from anatomy. Harrison, who studied both zoology and medicine in Baltimore and at the University of Bonn, became a professor first of anatomy at Johns Hopkins and then of comparative anatomy and zoology at Yale. His work at the intersection of cytology and embryology is best remembered for his contribution to tissue culture, which he developed building on earlier experiments of the aforementioned anatomist Hermann Braus.[27] Harrison, like Spemann, maintained a close relationship with German anatomy and the Anatomische Gesellschaft, even serving as the society's chairman in 1934/35, at a critical time when the international reputation of German science and medicine was challenged due to its wilful submission to Nazi ideology and policy – a role which he was willing to play, not only due to his long standing sympathies with Germany, but especially with German anatomy.[28]

### Discussion: Why have so few anatomists received the Nobel Prize?

During the first six decades of the 20th century, the Nobel Committee of physiology or medicine followed international research trends when awarding Nobel Prizes, starting with the work on infectious diseases (Emil von Behring, Ronald Ross, Niels Finsen, Robert Koch), awarding four scholars for the discovery of vitamins (Christiaan Eijkman, Frederick Hopkins, Henrik Dahm, Edward Doisy) and later shifting to biochemistry and molecular biology (Otto Warburg, Albert Szent-Györgi, Carl and Gerti Cori, Hans Krebs), especially genetics and DNA (Thomas Morgan, George Beadle, Edward Tantum, Joshua Lederberg, Severo Ochoa, Arthur Kornberg, Francis Crick, James Watson, Maurice Wilkins). In contrast, anatomy is a discipline with a weak track record in the Nobel context. Based on an analysis of Nobel Prize nominations and reports by the Nobel Committee, we suggest that two factors were important for this negative trend.

First, our overview shows that anatomy as a discipline was in decline in the Nobel context from an early stage. Some of the candidates were considered too old at the turn of the 20th Century. For example, Kölliker (b. 1817) was not deemed a suitable candidate for the Nobel Committee, looking for "visionary science" of the 20th century. The same can be said about Rudolf Virchow, who also did not become a strong contender for the prize, despite being nominated in 1901 and 1902, or Ernst Haeckel, nominated four times from 1902 to 1913. In several cases, the nominators for anatomists emphasized life-time-achievements and textbooks rather than single discoveries. This observation corresponds with the declining status of anatomy in other contexts, like German scientific academies. In addition, prominent scientists with strong ties to anatomy (e.g. Cécile and Oscar Vogt) were not enacted as neuroanatomists in the nominations in the first place, but rather as basic researchers.[29]

Second, during the first decades of the 20th century, the Nobel Committee favoured clinical researchers who aimed at solving research questions with far-reaching clinical implications "for the greatest benefit of mankind" (as stipulated by Alfred Nobel in his will). After 1940 the focus changed to emphasize basic researchers, but then hardly any anatomists were proposed. The cases of von Kölliker and Bargmann – different as their respective legacies may have been – display a general difficulty of anatomical research being acknowledged as a distinguished scientific contribution to medicine. The decoration of Golgi and Ramón y Cajal appears more like an exception to a rule or, rather, like a remnant of the dependence of medical knowledge on anatomical investigation. It can be considered a late acknowledgement of times gone by, that is, the unraveling of the last blank areas of the human form and structure, with little to no perceived potential for further innovation, especially in practical-clinical terms. The award of the Nobel Prize to Spemann in 1935 was, in contrast, the acknowledgment of a scientific potential that was only beginning to unfold. Both awards were, as their awardees, inextricably linked to anatomy, but it could be argued that the first pointed towards the past, while the second pointed towards the future – which meant, incidentally, away from anatomy.

## Acknowledgement

We are grateful to Sabine Hildebrandt, Harvard Medical School, for important comments.

## Notes

1 Nils Hansson, Giacomo Padrini, Friedrich H. Moll, Thorsten Halling, Carsten Timmermann, "Why so Few Nobel Prizes for Cancer Researchers? An analysis of Nobel Prize Nominations for German Physicians with a Focus on Ernst von Leyden and Karl Heinrich Bauer,“ *Journal of Cancer. Research and Clinical Oncology* 147 (2021), 2547–2553.

2 Hansson Nils, Thorsten Halling and Heinder Fangerau, eds., *Attributing Excellence in Medicine: The History of the Nobel Prize* (Leiden: Brill, 2019).

3 Rudolf Virchow, "Morgagni und der anatomische Gedanke," *Berliner Klinische Wochenschrift* 31 (1894), 345–350.

4 Lynn K. Nyhart, *Biology takes Form: Animal Morphology and the German Universities, 1800–1900* (Chicago: University of Chicago Press, 1995).

5 "Anatomische Gesellschaft," edited by Karl von Bardeleben, *Anatomischer Anzeiger* 1 (1886), 236–237.

6 Thorsten Halling, Nils Hansson and Heiner Fangerau, "'Prisvärdig' Forschung? Wilhelm Roux und sein Programm der Entwicklungsmechanik," *Berichte zur Wissenschaftsgeschichte* 41 (2018), 73–97.

7 N.N., "Albert von Kölliker (1817–1905) Würzburger Histologist," *JAMA* 206 (1968), 2111.

8 C. Mörgeli, „Kölliker, Albert,“ in *Historisches Lexikon der Schweiz* (HLS) (2007). Online: https://hls-dhs-dss.ch/de/articles/014436/2007-08-23/ (accessed December 28, 2022).

9 Bardeleben, "Anatomische Gesellschaft."

10 R. Hildebrand, "Der Würzburger Anatom Albert von Koelliker in seiner Beziehung zu Camillo Golgi und Santiago Ramón y Cajal," *Sudhoffs Archiv* 73 (1989), 145–155.

11 Pedro Mestres-Ventura, "Albert von Kölliker, Santiago Ramón y Cajal and Camillo Golgi, the Main Protagonists in the Neuron Theory Debate," *European Journal of Anatomie* 23 (2019), 5–14.

12 Hans Krebs, "Wie ich aus Deutschland vertrieben wurde," *Medizinhistorisches Journal* 15 (1980), 357–377.

13 Sabine Hildebrandt, "Wolfgang Bargmann (1906-1978) and Heinrich von Hayek (1900-1969): Careers in Anatomy Continuing through German National Socialism to Postwar Leadership," *Annals of Anatomy* 195 (2013), 283–295.

14 J. Pritchard, "In memoriam: Wolfgang Bargmann," *Journal of Anatomy* 128 (1979), 407.

15 H. Eggeling, ed., *Verhandlungen der Anatomischen Gesellschaft.* 42 (1934).

16 C. Andree, T. Schwarz and R. Fölster-Holst, "Universitäts-Hautklinik Kiel der Christian-Albrechts-Universität zu Kiel 1902–2016," *Aktuelle Dermatologie* 42 (2016), 13–19.

17 Nils Hansson and Sabine Hildebrandt, "Swedish-German Contacts in the Field of Anatomy 1930-1950: Gösta Häggqvist and Hermann Stieve," *Annals of Anatomy* 196 (2014), 259–267.

18 B. Scharrer, "Wolfgang Bargman. 1906–1978," *The Anatomical Record* 195 (1979), 145–147; W. Bargmann and E. Scharrer, "The Site of the Hormones of the Posterior Pituitary," *American Journal of Science* 39 (1959), 255–259.

19 Christina Brandt, "Vitalism, Holism, and Metaphorical Dynamics of Hans Spemann's 'Organizer' in the Interwar Period," *Journal for the History of Biology* 55 (2022), 285–320.

20 Eggeling, ed., *Verhandlungen der Anatomischen Gesellschaft* 42 (1934).

21 Hans Spemann, *Forschung und Leben* (Stuttgart: J. Engelhorn, 1943).

22 H. Braus, "Versuch einer experimentellen Morphologie," *Münchener Medizinische Wochenschrift* 50 (1903), 2076–2077.

23 Peter E. Fäßler, *Hans Spemann 1869–1941. Experimentelle Forschung im Spannungsfeld von Empirie und Theorie: Ein Beitrag zur Geschichte der Entwicklungsphysiologie zu Beginn des 20. Jahrhunderts* (Berlin, Heidelberg: Springer 1997).

24 H. Braus, "Verlagsbuchhandlung Julius Springer. Schlußwort zum letzten Heft des Archivs für mikroskopische Anatomie vor seiner Vereinigung mit dem Archiv für Entwicklungsmechanik," *Archiv für Mikroskopische Anatomie* 97 (1923), n.p.

25 Hans Spemann, "Hermann Braus," *Wilhelm Roux' Archiv für Entwicklungsmechanik der Organismen* 106 (1925), I–XXII.

26 A. Benninghoff, "Die biologische Feldtheorie," *Sitzungsberichte Heidelbergischer Akademie der Wissenschaften. Mathematisch-Naturwissenschaftliche Klasse* 4 (1942).

27 Sabine Brauckmann, "Networks of Tissue Knowledge. 1910–1960," *Bulletin d'histoire et d'épistémologie des sciences de la vie* 13 (2006), 37–56.

28 Florian Steger and Mathias Schütz, "Anatomiegeschichte – Akademiegeschichte. Zur Entwicklung der Wissenschaftsakademien im Dritten Reich am Beispiel der Anatomie," *Acta historica Leopoldina* 64 (2014), 259–288.

29 Nils Hansson, *Wie man keinen Nobelpreis gewinnt – Die verkannten Genies der Medizingeschichte* (München: Gräfe & Unzer, 2023).

## Bibliography

Andree, C., T. Schwarz and R. Fölster-Holst. "Universitäts-Hautklinik Kiel der Christian-Albrechts-Universität zu Kiel 1902–2016." *Aktuelle Dermatologie* 42 (2016): 13–19.

Bardeleben, K. von, ed. "Anatomische Gesellschaft," *Anatomischer Anzeiger* 1 (1886): 236–237.

Bardeleben, K. von, ed. "Anatomische Gesellschaft," *Verhandlungen der Anatomischen Gesellschaft* 8 (1894).

Bargmann, W. and E. Scharrer. "The Site of the Hormones of the Posterior Pituitary," *American Journal of Science* 39 (1951): 255–259.

Benninghoff, A. "Die biologische Feldtheorie," *Sitzungsberichte Heidelbergischer Akademie der Wissenschaften. Mathematisch-Naturwissenschaftliche Klasse* 4 (1942).

Brandt, Chistina. "Vitalism, Holism, and Metaphorical Dynamics of Hans Spemann's 'Organizer' in the Interwar Period," *Journal for the Historie of Biologie* 55 (2022): 285–320.

Brauckmann, Sabine. "Networks of Tissue Knowledge. 1910–1960," *Bulletin d'histoire et d'épistémologie des sciences de la vie* 13 (2006): 37–56.

Braus, H. "Verlagsbuchhandlung Julius Springer. Schlußwort zum letzten Heft des Archivs für mikroskopische Anatomie vor seiner Vereinigung mit dem Archiv für Entwicklungsmechanik," *Archiv für Mikroskopische Anatomie* 97 (1923), n.p.

Braus, H. "Versuch einer experimentellen Morphologie," *Münchener Medizinische Wochenschrift* 50 (1903): 2076–2077.

Eggeling, H., ed. *Verhandlungen der Anatomischen Gesellschaft* 42 (1934).

Fäßler, Peter E. *Hans Spemann 1869–1941. Experimentelle Forschung im Spannungsfeld von Empirie und Theorie: Ein Beitrag zur Geschichte der Entwicklungsphysiologie zu Beginn des 20. Jahrhunderts.* Berlin, Heidelberg: Springer, 1997.

Halling, Thorsten, Nils Hansson and Heiner Fangerau. "'Prisvärdig' Forschung? Wilhelm Roux und sein Programm der Entwicklungsmechanik," *Berichte zur Wissenschaftsgeschichte* 41 (2018): 73–97.

Hansson Nils and Sabine Hildebrandt. "Swedish-German Contacts in the Field of Anatomy 1930-1950: Gösta Häggqvist and Hermann Stieve," *Annals of Anatomy* 196 (2014), 259–267.

Hansson Nils. *Wie man keinen Nobelpreis gewinnt – Die verkannten Genies der Medizingeschichte.* München: Gräfe & Unzer, 2023.

Hansson Nils, Thorsten Halling and Heinder Fangerau, eds. *Attributing Excellence in Medicine: The History of the Nobel Prize.* Leiden: Brill, 2019.

Hansson, Nils, Giacomi Padrini, Friedrich H. Moll, Thorsten Halling and Carsten Timmermann, "Why so Few Nobel Prizes for Cancer Researchers? An Analysis of Nobel Prize Nominations for German Physicians with a Focus on Ernst von Leyden and Karl Heinrich Bauer," *Journal of Cancer Research and Clinical Oncology* 147 (2021): 2547–2553.

Hildebrand, R. "Der Würzburger Anatom Albert von Koelliker in seiner Beziehung zu Camillo Golgi und Santiago Ramón y Cajal," *Sudhoffs Archiv* 73 (1989): 145–155.

Hildebrandt, Sabine. "Wolfgang Bargmann (1906-1978) and Heinrich von Hayek (1900-1969): careers in anatomy continuing through German National Socialism to postwar leadership," *Annals of Anatomy* 195 (2013): 283–295.

Krebs, H. "Wie ich aus Deutschland vertrieben wurde," *Medizinhistorisches Journal* 15 (1980): 357–377.

Mestres-Ventura, Pedro. "Albert von Kölliker, Santiago Ramón y Cajal and Camillo Golgi, the Main Protagonists in the Neuron Theory debate," *European Journal of Anatomie* 23 (2019): 5–14.

Mörgeli, C. "Kölliker, Albert." In *Historisches Lexikon der Schweiz* (HLS), 2007. https://hls-dhs-dss.ch/de/articles/014436/2007-08-23/ (accessed December 28, 2022).

N.N. "Albert von Kölliker (1817–1905) Würzburger Histologist," *JAMA* 206 (1968), 2111.

Nyhart, Lynn K. *Biology takes Form: Animal Morphology and the German Universities, 1800–1900*, Chicago; University of Chicago Press, 1995.

Pritchard, J. "In memoriam: Wolfgang Bargmann," *Journal of Anatomie* 128 (1979): 407.

Scharrer, B. "Wolfgang Bargman. 1906–1978," *The Anatomical Record* 195 (1979): 145–147.

Spemann, H. "Hermann Braus," *Wilhelm Roux' Archiv für Entwicklungsmechanik der Organismen*, 106 (1925): I–XXII.

Spemann, Hans. *Forschung und Leben.* Stuttgart: J. Engelhorn, 1943.

Steger, Florian and Mathias Schütz. "Anatomiegeschichte – Akademiegeschichte. Zur Entwicklung der Wissenschaftsakademien im Dritten Reich am Beispiel der Anatomie," *Acta Historica Leopoldina* 64 (2014): 259–288.

Virchow, Rudolf. "Morgagni und der anatomische Gedanke," *Berliner Klinische Wochenschrift* 31 (1894), 345–350.

CHAPTER ELEVEN

# Physicians as Candidates for the Nobel Peace Prize

*Leander Scheel & Nils Hansson*

**Abstract**

When considering physicians who were nominated for the Nobel Prize, focusing solely on the Nobel Prize in Physiology or Medicine is insufficient. Doctors have also been put forward for other prize categories, as recent accounts have shown for the prize categories chemistry and literature. Compared to other Nobel Prize categories, there are surprisingly few scientific publications dealing with the influence of physicians on the Nobel Peace Prize. Which physicians were nominated for the Nobel Peace Prize? To what extent was medicine a topic in the nominations and Nobel Committee evaluations?

**Keywords:** Medicine and Peace, Nobel Peace Prize, Albert Schweitzer, Josué de Castro, Ludwig Zamenhof

When considering physicians nominated for the Nobel Prize, focusing solely on the Nobel Prize in Physiology or Medicine is insufficient. Doctors have also been put forward for other prize categories, as recent accounts have shown for the prize categories chemistry and literature.[1] Compared to other Nobel Prize categories, however, there are surprisingly few scientific publications dealing with the influence of physicians on the Nobel Peace Prize. Which physicians were nominated for the Nobel Peace Prize? To what extent was medicine a topic in the nominations and Nobel Committee evaluations?

According to Alfred Nobel's will, the Nobel Peace Prize should be awarded to those who "have done the most or best to advance fellowship among nations, the abolition or reduction of standing armies, and the establishment and promotion of peace congresses."[2] Drawing on nominations and assessments by the Nobel Committee, this commentary provides an overview of physicians as Nobel nominees for the Peace Prize from 1901 to 1973. This paper, complemented by the description of a few case-studies below, might serve as material for future research on nomination processes.

Following a fifty-year embargo, researchers can apply to study the nomination letters and the Nobel Committee evaluations, so that we now (December 2023) have access to 4690 nominations that were submitted for the Nobel Peace Prize from 1901 to 1973. During this period, 25 physicians (all male) were nominated at least once. Only two of them, Lord (John) Boyd Orr (1949) and Albert Schweitzer (1952), were eventually awarded the Nobel Peace Prize. The overview given in table 1 (Physicians Nominated for the Nobel Peace Prize 1901–1973) shows that only a few physicians were portrayed as promoters of world-peace because of their background or activities in medicine. One example is Paul Dudley White (1886–1973), an American cardiologist and founder of the American Heart Association, nominated for diplomatic efforts to build bridges between US and Soviet scientists during the Cold War.[3] The table shows what was suggested earlier, namely that the number of nominations alone does not seem to increase the chances of receiving the prize (only two nominations for Orr).[4] Instead, the reasons behind the decisions can only be reconstructed by studying the original sources. In this paper, we will show why three of the candidates (Louis Lazare Zamenhof, Josué de Castro, and Albert Schweitzer) were nominated for the Nobel Peace Prize, by whom, when, and the reasons why they did (not) receive the prize. All three doctors made it onto the Nobel Committee's shortlist. Zamenhof and de Castro were each nominated once by a Nobel laureate: Charles Robert Richet, a laureate for Physiology or Medicine, and the aforementioned Lord (John) Boyd Orr.

*Table 1: Physicians Nominated for the Nobel Peace Prize 1901-1973*

| **Number and Years of Nominations for Peace** | **Nominated Doctor (Country)** | **Reason for Nomination** |
|---|---|---|
| 1 time, 1901 | Arthur Mühlberger (DE) | for his comprehensive thoughts on war and peace, particularly expressed in his 1899 book on Proudhon. |
| 2 times, 1907, 1908 | C S Leadbetter (GB) | / |
| 14 times, 1907, 1909, 4x1910, 3x1913, 1914, 1915, 2x1916, 1917 | Louis Lazare Zamenhof (PL) | for Esperanto |
| 4 times, 3x1923, 1924 | Frédéric Ferrière (CH) | founder and leader of "Agence internationale de secours et de renseignement en faveur des prisonniers de guerre" |
| 1 time, 1916 | Enrico de Gama Coelho | For his writings on women in war |
| 5 times, 5x1928 | Auguste-Henri Forel (CH) | / |

| | | |
|---|---|---|
| 30 times, 1930, 5x1931, 1932, 2x1933, 1934, 4x1935, 6x1936, 3x1938, 1939, 3x1950, 2x1952, 1953 | Albert Schweitzer (FR)<br>Nobel Peace Prize 1952 | Founder Lambaréné Hospital, Gabon – then a province of French Equatorial Africa |
| 1 time, 1932 | Christian F Heerfordt (DK) | for his advocacy of a "United States of Europe" |
| 1 time, 1937 | N A Nilsson (SE) | / |
| 1 time, 1937 | G Saint-Paul (FR) | / |
| 2 times, 1947, 1949 | Lord (John) Boyd Orr von Brechin (GB)<br>Nobel Peace Prize 1949 | for his contributions to food policy during WWII, leadership in the UN Food and Agriculture Organization, efforts to prevent global famine, and roles in peace organizations. |
| 3 times, 1952, 1954, 1958 | Lorenzo Fernandez Rodriguez (CL) | for his literary work on social-political and international issues |
| 8 times, 1901, 3x1951, 1952, 1953, 1954, 1955 | Louis Constant Vauthier (FR) | founder and the director of the International University Sanitarium in Leysin in Switzerland. |
| 8 times, 3x1953, 1963, 1964, 2x1965, 1970 | Josué de Castro (BR) | For his work on increase in population and access to food in his book "Geography of Hunger" |
| 4 times, 1953, 1969, 1970, 1971 | George Brock Chisholm (CA) | for his work as leader of the World Health Organisation |
| 5 times, 5x1970 | M C Davar (IN) | for reconciliation between Hindus and Muhamedans |
| 1 time, 1959 | Robert M Debré (FR) | for his work worldwide with child protection |
| 1 time, 1960 | Howard A Rusk (US) | for his rehabilitation work |
| 4 times, 1964, 1965, 1966, 1968 | Guido Guida (IT) | for founding the "Centro Internazionale Radio-Medico" |
| 4 times, 1964, 1965, 1966, 1968 | Abraham Vereide (US) | for his work towards the elimination of war from society, and efforts toward world-brotherhood and interracial fraternity trough his Christian prayer and action group |

| | | |
|---|---|---|
| 3 times, 1966, 1968, 1969 | Joaquín Sanz Gadea (ES) | for his humanitarian activities as a doctor in contributing to the peace in Stanleyville |
| 1 time, 1968 | Yoshio Koya (JP) | for his pioneering efforts as a world-renowned gynecologist that resulted in the decline in the birth rate and rate of induced abortions in Japan |
| 4 times, 4x1968 | Frans Hemerijckx (BE) | for his work for people suffering from leprosy in Africa and Asia |
| 3 times, 2x1969, 1970 | Paul Dudley White (US) | for peaceful understanding and cordial collaboration among cardiologists |
| 1 time, 1970 | François Duvalier (HT) | for his philosophy and his work for the poor masses of his country |

## Louis Lazare Zamenhof (1859-1917) – Ophthalmologist and Nobel Prize Nominee

Louis Lazare Zamenhof was a Jewish ophthalmologist from Poland. In 1887, he played a significant role in constructing Esperanto, an international language intended to facilitate worldwide understanding.[5,6] For this artificial language, Zamenhof was nominated for the Nobel Peace Prize 14 times between 1907 and 1917. Many of Zamenhof's nominators were themselves Esperanto enthusiasts, including Felix Moscheles, the first president of the London Esperanto Club.[,8,9]

In 1910, the Nobel Committee considered a possible Nobel Peace Prize for Zamenhof. At this point, six nominations had been submitted. Considering the arguments of the Nobel Committee, it is striking, first of all, how profound the Nobel Committee's background knowledge of Esperanto was. Esperanto, based on Romance, Germanic, and Slavic languages, had by then been widespread worldwide for more than twenty years. Thousands of people could speak it, books were being printed in Esperanto, 1500 societies, and 100 magazines were using Esperanto. The argument that Esperanto promoted peaceful cooperation among nations was adopted by the Nobel Committee from the nomination letters for Zamenhof. The Committee explained that before Esperanto, several languages had been created, such as Pidgin English or Russian-Norwegian, to facilitate trade. But the impact of Esperanto seemed more significant, as stated in the Nobel dossier. For example, other constructed languages such as Volapük had been replaced by Esperanto.

Zamenhof's endeavour to create Esperanto was inspired by the many languages spoken in his hometown Bialystok, where Russians, Poles and Germans lived together. Under the pseudonym "Dr. Esperanto," he published his first textbook in 1887: *Lingvo internacia.* Since then, his artificial language was named after the author's pseudonym: Esperanto. Nevertheless, the Nobel Committee criticized that not everyone who called themselves one of up to 1,000,000 (estimated 1908 at an Esperanto meeting) Esperantists could actually speak and read Esperanto, but only supported the idea of an international language. Furthermore, it needed improvement in terms of vocabulary, grammar, and writing style. In addition to these critical remarks, Walther Borgius, Vice President of the Berlin Local Group of the German Esperanto Society, wrote in his book "Why I Left Esperanto" (1908) that Esperanto had in no way created new avenues for understanding among people.[10]

The Nobel Committee feared that Esperanto would suffer the same fate as Volapük. In the conclusion of the Nobel Committee's considerations on Zamenhof, the members indicated that the idea of a common language used by a large part of the world's population was indeed important. However, whether an artificial language like Esperanto would ever be used for this purpose was questionable. According to the Committee, English would be better suited and that was the main reason why Zamenhof's nomination was not endorsed by the jury members.[11]

## Josué Apolônio de Castro (1908-1973) – Physician, Nutritionist, and Nobel Prize Nominee

In 1929, at the age of 21, de Castro completed his medical degree in Rio de Janeiro. His interests were not limited to medicine and nutrition sciences. He became well-known as geographer, politician, diplomat, and international activist against world hunger. His most famous books, *Geography of Hunger* (1947) and *Geopolitics of Hunger* (1951), demonstrating the significance of vitamins, proteins, and mineral salts in nutrition, were translated into more than 25 languages.[12]

De Castro held various academic and political positions throughout his life, for example assistant professor of physiology (1932), professor of cultural geography, associate dean of the School of Philosophy and Social Sciences in Recife (1933), anthropologist at the University of Brazil (1935), and professor of nutritional sciences. He also served as a delegate at several conferences and led the Department of the Ministry of Economics (National Technical Bureau of Food 1942-44) for coordinating nutritional issues. De Castro was elected as a member of the Brazilian Chamber of Deputies for eight years (1954 to 1962). In 1962, he was Brazil's ambassador to the UN in Geneva and Brazil's delegate to the Disarmament Conference in Geneva in 1963. In 1957, he was elected president of the Council of the FAO (Food and Agriculture Organization of the United Nations). He argued for greatly reducing countries' military spending, even towards total disarmament. He also demanded a ban on nuclear weapon testing to provide financial resources for the economic and social development of a peaceful world.

De Castro was nominated a total of eight times for the Nobel Peace Prize between 1953 and 1970, reaching the Nobel Committee's shortlist twice, in 1963 and 1970. In 1953, he was nominated by two professors (Eremildo Luiz Viana and Victor Leal) from the Universidade do Brasil, where Josué de Castro also worked.[13] He had personal ties with other nominators as well. He had corresponded with his nominators Richard Acland (British MP), Aneurin Bevan (former British Minister of Health), Gilbert McAllister (British MP), and Lord Silkin (British MP). He was surprised by the kindness of the four Britons who nominated him, as he had frequently criticized Britain and its colonial policies.[14]

During the Nobel Committee's deliberations, de Castro's work *Geography of Hunger* was cited as the primary argument his nominators made for his nomination for the Nobel Peace Prize. The Nobel Committee described it as a polemic review. When he was shortlisted for the Nobel Peace Prize a second time in 1970, his work was dismissed as a book without scientific ambitions intended for a broad readership. These arguments can be found in the advisory reports of the Nobel Committee. In de Castro's eyes, hunger was not just a lack of calories. He specifically highlighted the impacts of hidden hunger, such as the lack of protein, minerals, or vitamins, resulting in beriberi, pellagra, scurvy, xerophthalmia, rickets, osteomalacia, endemic goiter, anemia, and other detrimental influences on human physical and mental health.[15] He blamed colonialism and imperialist exploitation, for instance, through private land ownership by large landowners and monocultures, for the hunger in his home country Brazil, among other places. The production of food was too unvaried, and unessential food was being grown. He was convinced that overpopulation did not necessarily lead to poverty and inadequate nutrition. In his opinion, the population in contrast would drastically decrease if nutritional conditions were improved. An artificial limitation of the population would not be necessary. Underpopulation would be the result. Here, de Castro applied a rat experiment to humans, which shows that a protein-rich diet, including animal products, reduces the number of children. He claimed to have evidence for this, consisting of tables of protein content in the diet and the birth rate in different countries. Countries with a protein-rich diet and a high birth rate are not mentioned by de Castro. The Nobel Committee criticized de Castro for not questioning the connection between fertility and nutrition in terms of proteins.[16] In addition, they argued that his ideas lacked an impact in practice.[17]

### Albert Schweitzer (1875-1965), Nobel Laureate 1952

The German-French doctor, theologian, philosopher, musician, and peace activist is known to this day for his ethics of reverence for life.[18] Albert Schweitzer was nominated once for the Literature Prize in 1952, and an exceptional thirty times for the Nobel Peace Prize between 1930 and 1952. Schweitzer's nominators included such well-known figures as the Swedish historian Nils Ahnlund (1889-1957), the American historian Sidney Bradshaw Fay (1876-1967), the American philosopher Robert Maynard Hutchins (1899-1977), and the Norwegian politician and philosopher Kjell Bondevik (1901-1983). Interestingly, Schweitzer's nominators were frequently philosophers, historians, pastors, lawyers, and politicians, who all except three of the nominators came from Europe. In contrast to Josué de Castro and L.L. Zamenhof, none of his nominators was a Nobel laureate.[19,20]

Albert Schweitzer was nominated for his humanitarian work in Gabon, a former French colony, where he built the Lambaréné Hospital in 1913 with his wife after completing his medical training. Moreover, Schweitzer engaged deeply in philosophy. One of his most important books was *Philosophy of Civilization*, in which the concept of reverence for life was developed as an ethical concept, which excludes any legitimization for warlike actions. Additionally, he was praised for strengthening German-French friendship. His nominators repeatedly emphasized his selfless dedication in the Lambaréné Hospital in Gabon. They noted that Schweitzer explained in his book *On the Edge of the Primeval Forest* (1929) that the West must offset the misery and injustice caused by colonialism and imperialism through solidarity and aid. The best sources to judge his activities were, according to the committee, the books *Out of My Life and Thought* (1931), *On the Edge of the Primeval Forest* (1921), *Reports from Lambarene*(1924-1927), and *Albert Schweitzer – Thought and Action* (1950), in which the success of his medical activities is described. Schweitzer was motivated by his own philosophical concept of the ethics of "reverence for life."

The Nobel Committee mentioned three central points in favour of Schweitzer's candidacy. First, he had contributed to the reconciliation between peoples, especially between Germany and France. Second, he had worked for the reconciliation of different cultures, especially between white colonialists and people of colour in Africa. And third, through his lectures and books, he had promoted tolerance and mutual international understanding from a firm religious standpoint. However, one criticism of the Nobel Committee was that he probably did not actively contribute to the defusing of political conflicts. Schweitzer did not involve himself in current politics. He rejected National Socialism but did not participate in any anti-Nazi campaign. In the end, however, Albert Schweitzer received the Nobel Peace Prize in 1952 "for his altruism, reverence for life, and tireless humanitarian work which has helped making the idea of brotherhood between men and nations a living one."[21]

From 1973 to 2022, three more Nobel Peace Prizes were awarded to doctors and doctor-led organizations: 1. IPPNW (International Physicians for the Prevention of Nuclear War) in 1985, 2. Doctors Without Borders in 1999, and 3. Gynecologist Denis Mukwege for "his efforts to end the use of sexual violence as a weapon of war and armed conflict" in 2018. It can be assumed that more doctors were nominated for the Nobel Peace Prize since 1973. However, they are not yet known due to the 50-year embargo. Future studies will analyze further nominations for the Nobel Peace Prize to deepen our understanding of how medicine and science were enacted as vehicles for peace in general and how physicians and scientists were portrayed as promoters of world-peace in particular.

## Note on archival material

Norwegian Nobel Committee Archives, the Norwegian Nobel Institute, Oslo (nomination letters and internal communication within the Nobel Committee regarding the selection process) 1901-1973.

## Notes

1 Michael Pohar and Nils Hansson, "Between two stools? Pharmacologists nominated for Nobel prizes in 'physiology or medicine' and 'chemistry' 1901-1950 with a focus on John Jacob Abel (1857-1938)," *Naunyn Schmiedebergs Archives of Pharmacology*, 394(3), 503–513 (October, 2020); Nils Hansson, Peter M. Nilsson, Heiner Fangerau, Jonatan Wistrand, "The enactment of physician-authors in Nobel Prize nominations," *PLoS One*, 15(11):e0242498, (November 2020), https://doi.org/10.1371/journal.pone.0242498.

2 The Nobel Foundation and W. Odelberg, H. Schück, R. Sohlman, A. Österling, C.G. Bernhard, A. Westgren, M. Siegbahn, K. Siegbahn, A. Schou, N.K. Ståhle, eds., *Nobel: The man and his prizes*, 3rd ed. (1972), (New York, London, Amsterdam: American Elsevier Publishing Company, 1950), 10.

3 Nils Hansson, Thomas Schlich, "Performing Excellence: Nobel Prize nomination networks in North America," *Notes and Records of The Royal Society Journal of the History of Science*, (December 15, 2021).

4 N. Hansson, U. Schagen, "In Stockholm hatte man offenbar irgendwelche Gegenbewegung– Ferdinand Sauerbruch (1875-1951) und der Nobelpreis," *NTM Zeitschrift für Geschichte der Wissenschaften, Technik und Medizin* 22(3), (2014), 133–161.

5 cf. Doris Bensimon, "Archives de sciences sociales des religions," JSTOR, No. 96 (October-December, 1996), 140–141.

6 cf. Andreas Künzli, *L.L. Zamenhof (1859-1917): Esperanto, Hellenismus (Homaranismus) und die "jüdische Frage" in Ost- und Westeuropa*, (Wiesbaden: Harrassowitz Verlag, 2010), 11.

7 C.S. Alihusain, "Erinnerung an Felix Moscheles (1833-1917)," *peacepalacelibrary*, (January 11, 2018) https://peacepalacelibrary.nl/blog/2018/remembering-felix-moscheles-1833-1917 (accessed May 13, 2023).

8 Gaston, Moch, Foreword in "Esperanto language committee," *Esperantista Dokumentaro*, (Paris: Esperanto Central Office, 1910).

9 L. Kökény, V. Bleier, I. Sirjaev, *Enciklopedio de Esperanto*, (Budapest: Hungara Esperanto-Asocio, 1986), 69–82.

10 Walter Bergius, *Warum ich Esperanto verließ*, (Berlin: Liebheit & Thiesen, 1908) 69.

11 Norwegian Nobel Committee Archives, (Oslo: The Norwegian Nobel Institute, 1910).

12 Rosana, Magalhães, *Fome: Uma (re) leitura de Josué dee Castro,* (Rio de Janeiro: Editora Fiocruz, 1997), 25 ff.

13 cf. Marieta de Moraes Ferreira, "Anpuh XXV Simpósio nacional de história Fortaleza," ANPUH, (2009) 3, 5 ff., https://anpuh.org.br/uploads/anais-simposios/pdf/2019-01/1548772007_ae7186e-34a26edbof08203184c4d1f08.pdf (accessed May 13, 2023)

14 Archie Davies, "Josué de Castro´s Geografia Combatente and the political ecology of hunger," PhD diss. in Geography (London: King's College, 2019) 124, https://www.academia.edu/80927484/Josue_de_Castros_Geografia_Combatente_and_the_political_ecology_of_hunger?f_ri=3126 (accessed May 13, 2023).

15 Josué de Castro, *Geography of Hunger* (Boston: Brown and Company, 1952), 11.

16 de Castro, *Geography of Hunger*, X.

17 Norwegian Nobel Committee Archives, (Oslo: The Norwegian Nobel Institute, 1963, 1970).

18 Tizian Zumthurm, *Practicing Biomedicine at the Albert Schweitzer Hospital 1913-1965*, (Leiden/Boston: Brill, 2020).

19 THE NOBEL PRIZE, "Nomination Archive," https://www.nobelprize.org/nomination/peace/, (accessed May 13, 2023).

20 THE NOBEL PRIZE, "Albert Schweitzer," https://www.nobelprize.org/prizes/peace/1952/schweitzer/biographical/, Nomination Archive, (accessed May 13, 2023).

21 Norwegian Nobel Committee Archives, (Oslo: The Norwegian Nobel Institute, 1950, 1952).

## Bibliography

Alihusain, C.S. "Erinnerung an Felix Moscheles (1833-1917)." *peacepalacelibrary*, (January 11, 2018) https://peacepalacelibrary.nl/blog/2018/remembering-felix-moscheles-1833-1917 (accessed May 13, 2023).

Bensimon, Doris. "Archives de sciences sociales des religions," *JSTOR* no. 96 (December 1996), 140–141.

Bergius, Walter. *Warum ich Esperanto verließ.* Berlin: Liebheit & Thiesen, 1908.

Davies, Archie. 'Josué de Castro´s Geografia Combatente and the political ecology of hunger' (PhD diss., King`s College London, 2019), 124. https://www.academia.edu/80927484/Josue_de_Castros_Geografia_Combatente_and_the_political_ecology_of_hunger?f_ri=3126 (accessed May 13, 2023).

de Castro, Josué. *Geography of Hunger.* Boston: Brown and Company, 1952.

de Moraes Ferreira, Marieta. "Anpuh XXV Simpósio nacional de história Fortaleza," *ANPUH* (2009): 5 ff., https://anpuh.org.br/uploads/anais-simposios/pdf/2019-01/1548772007_ae7186e34a26edbof-08203184c4d1f08.pdf (accessed May 13, 2023).

Hansson, Nils and Thomas Schlich. "Performing Excellence: Nobel Prize nomination networks in North America." *Notes and Records of the Royal Society Journal of the History of Science* (2024): 2. https://doi.org/10.1098/rsnr.2021.0052.

Hansson, Nils and Udo Schagen. "In Stockholm hatte man offenbar irgendwelche Gegenbewegung– Ferdinand Sauerbruch (1875-1951) und der Nobelpreis." *NTM Zeitschrift für Geschichte der Wissenschaften, Technik und Medizin* (2014): 133–161.

Hansson, Nils, Peter M. Nilsson, Heiner Fangerau and Jonatan Wistrand. "The enactment of physician-authors in Nobel Prize nominations," *PLoS One* (November 2020), https://doi.org/10.1371/journal.pone.0242498.

Kökény, L., V. Bleier and I. Sirjaev. *Enciklopedio de Esperanto.* Budapest: Hungara Esperanto-Asocio, 1986.

Künzli, Andreas. *L.L. Zamenhof (1859-1917): Esperanto, Hellenismus (Homaranismus) und die "jüdische Frage" in Ost- und Westeuropa.* Wiesbaden: Harrassowitz Verlag, 2010.

Magalhães, Rosana. *Fome: Uma (re) leitura de Josué dee Castro.* Rio de Janeiro: Editora FIOCRUZ, 1997.

Moch, Gaston. "Esperanto language committee," *Esperantista Dokumentaro* Paris: Esperanto central office, 1910.

Norwegian Nobel Committee Archives. Oslo: The Norwegian Nobel Institute 1910, 1950, 1952, 1963, 1970.

Pohar, Michael and Nils Hansson. "Between two stools? Pharmacologists nominated for Nobel prizes in 'physiology or medicine' and 'chemistry' 1901-1950 with a focus on John Jacob Abel (1857-1938)." *Naunyn Schmiedebergs Archives of Pharmacology* (October 2020), 503–513.

The Nobel Foundation and W. Odelberg, H. Schück, R. Sohlman, A. Österling, C.G. Bernhard, A.Westgren, M. Siegbahn, K. Siegbahn, A. Schou, N.K. Ståhle. *Nobel: The man and his prizes, 3. Edition (1972).* New York, London, Amsterdam: American Elsevier Publishing Company, 1950.

THE NOBEL PRIZE. "Albert Schweitzer," https://www.nobelprize.org/prizes/peace/1952/schweitzer/biographical/, Nomination Archive (accessed May 13, 2023).

THE NOBEL PRIZE. "Nomination Archive," https://www.nobelprize.org/nomination/peace/ (accessed May 13, 2023).

Zumthurm, Tizian. *Practicing Biomedicine at the Albert Schweitzer Hospital 1913-1965*, Leiden/Boston: Brill, 2020.

# About the Authors

**Daniela Angetter-Pfeiffer**, born in Vienna in 1971, studied History and German Philology with a focus on medical and military history at Vienna University; Dr. phil. 1995. 1996-2001 she was employed at the Institute for the History of Medicine at the University of Vienna. In 2001 she changed to the Austrian Academy of Sciences, specializing in history of medicine, military history, and history of science. In her spare time, she works as a volunteer emergency paramedic with the Red Cross.

**Rob van den Berg** is a science historian at Leiden University. He graduated in 2021 on a biography of the first Nobel Prize laureate in chemistry, J.H. van 't Hoff, *Een gedreven buitenstaander*.

**Annelie Drakman** obtained her PhD in the History of Science and Ideas at Uppsala University in 2018. Her work on bloodletting received international recognition. She has been a visiting fellow at King's College London, New York University and Harvard University, and was the PI of the project *Joy in Science—the Function of Positive Emotions in Autobiographies by Nobel Laureates in Physics*. Currently, she works at Stockholm University.

**Christiaan Engberts** has published on conceptions of scholarly virtue among nineteenth-century scholars. He is currently working as a postdoctoral researcher on a transnational history of the University in Leuven in the context of the upcoming celebration of its first university in 1425.

**Nils Hansson** is Professor of the History, Philosophy and Ethics of Medicine at the Heinrich-Heine-University in Düsseldorf, Germany. His research interests include recognition in the sciences, research ethics, and medical history in the Baltic Sea region.

**Jelmer Heeren** studied history, theology and philosophy. He is currently a PhD candidate at the Faculty of Religion and Theology, VU Amsterdam, exploring the meaning of the work of Dutch historian of science Reijer Hooykaas (1906-1994). The project is co-financed by the VUvereniging.

**Gustav Källstrand** is an historian of science and the Senior Curator at the Nobel Prize Museum in Stockholm. He received his PhD from the Department of Culture and Society at Linköping University and has written books and papers on the history of the Nobel Prize from various perspectives ranging from the public image of the prize and the laureates to the inner workings of the prize awarding committees.

**Louise Lagarde** is a master's student in history at Leiden University. She is interested in the history of scientific fame and Nobel Prize laureates and had the opportunity to work on research for this paper during an internship at the Rijksmuseum Boerhaave. This research also formed the basis of her master's thesis on Hendrik Antoon Lorentz.

**Daniela Link** is a literary scientist and postdoc at the Institute for History, Philosophy and Ethics of Medicine at Heinrich Heine University Düsseldorf and works in the field of Health Humanities. Her research focuses on cultural autism studies and the literary representation and perception of the Nobel Prize.

**Ad Maas** is curator of Rijksmuseum Boerhaave, the Dutch National Museum of the History of Science and Medicine, and Professor in Museological Aspects of the Natural Sciences at Leiden University.

**Giacomo Padrini** is a student researcher at the Department for the History, Philosophy, and Ethics of Medicine at Heinrich Heine University in Düsseldorf since 2019. His main interests are recognition, awards and prizes in science.

**Leander Scheel** is a medical student and doctoral candidate at the Department for the History, Philosophy, and Ethics of Medicine at Heinrich Heine University in Düsseldorf. His research focuses on prizes in medicine.

**Mathias Schütz** is a historian of medicine at the Institute of Ethics, History and Theory of Medicine at LMU Munich. His research focuses on the history of medical and bioethics, of Nazi medicine and, in particular, the intersections of these topics with the history of anatomy in the 20th century.

**Andreas Winkelmann** is professor of anatomy at Brandenburg Medical School in Neuruppin, Germany, since 2015. A medical doctor by training, he holds an additional MSc. degree in medical anthropology. He has researched the history and ethics of anatomy, including the history of the Anatomische Gesellschaft and of anatomy during the "Third Reich," and has conducted provenance research on colonial skeletal collections leading to several repatriations of human remains.

# Index of Names

Zeitfracht Medien GmbH
Ferdinand-Jühlke-Straße 7
99095 Erfurt, Deutschland
produktsicherheit@kolibri360.de